Contents

1. Introduction (from Article 313)
2. Article 275. Planet X debris field impacting earth
3. Article 282. Earth in upheaval: magma rising from beneath
4. Article 291. The sun disappears: day turns into night
5. Article 304. Can Planet X cause asteroid type impacts?
6. Article 288. Image of the center of the Milky Way Galaxy reveals how the universe works
7. Article 316. A rogue planet with a magnetic field 200 times stronger than Jupiter's
8. Article 317. Planet X System: Brown and Black Dwarfs with high magnetic field
9. Article 323. Planet X System Gravity: why has the earth not been destroyed?
10. Article 324. Planet X causes the sun to be darkened
11. Article 336. Stellar nebular cloud structure
12. Article 337. Heat and gravitational photon energy
13. Article 338. The Planet X effect: heating and ionization in contact regions
14. Article 339. Planet X and the Interstellar Medium: can we leave the Solar System?
15. Article 347. Gravity wave on Venus suggests Planet X presence
16. Article 348. Venus bulge: gravitational waves and hollow planets
17. Article 335. Biological and Ecological weapons in use against us
18. Article 342. The enemy: Why are bio and eco-weapons being used against us?

Books previously published

Book 1: Planet X: the awakening is now.
Book 2: The Planet X Report 2017: Photographic Evidence.
Book 3: Planet X Revealed Gravity and Light.
Book 4: The Sun Simulator
Book 5: Chemtrails: The Silent Killer.
Book 6: Planet X Physicist Articles: Part 1
Book 7: Planet X: The effects on the Earth and the Sun

Chapter 1

Introduction

I recently was asked in a comment why I think other astrophysicists do not support me. I think this is a very interesting and valid question. Why would any astrophysicist ignore the irrefutable evidence that there is a system of objects in the Solar System, which are found in the Sun's corona, and that is affecting our planet in unprecedented and cataclysmic ways? Evidence to their presence is undeniable and some of this evidence is shown below.

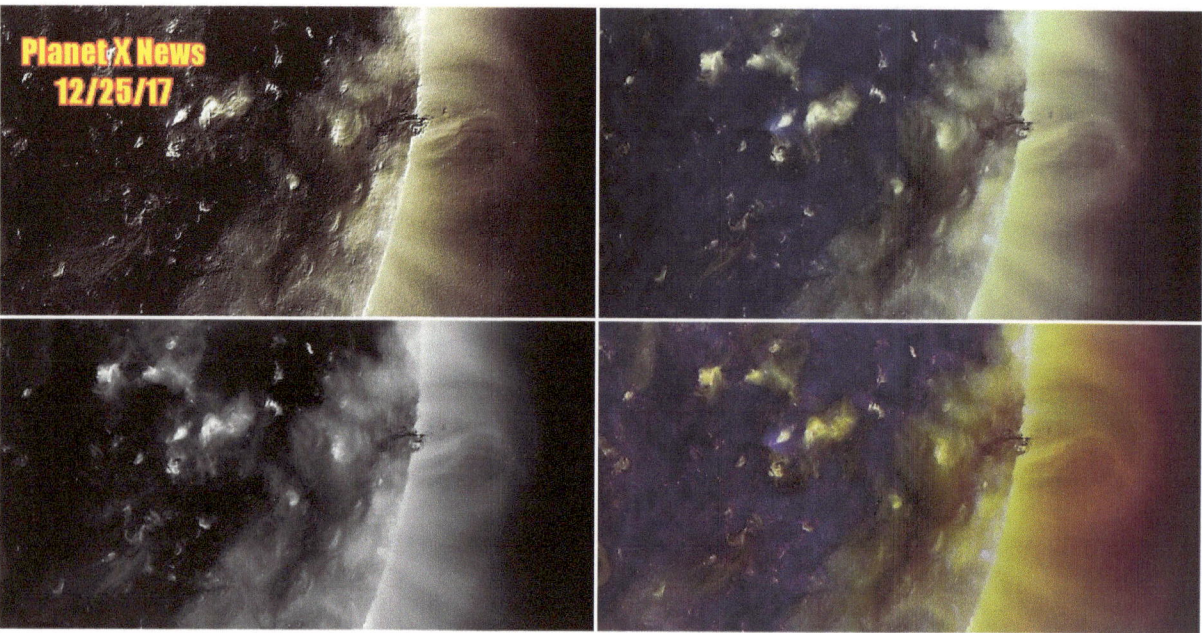

Figure 1.1. A Planet X Object or Stellar Core appears in the Sun's corona. The object is striped and a size comparison with the Sun reveals that it is about 4 times larger than the earth.

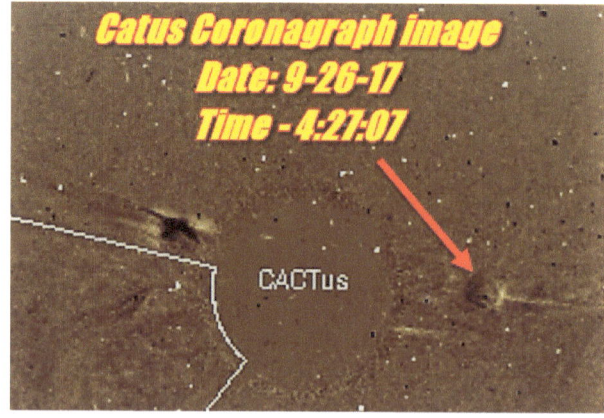

Figure 1.2. A spherical object moves away from the Sun within CME material in this CACTus image.

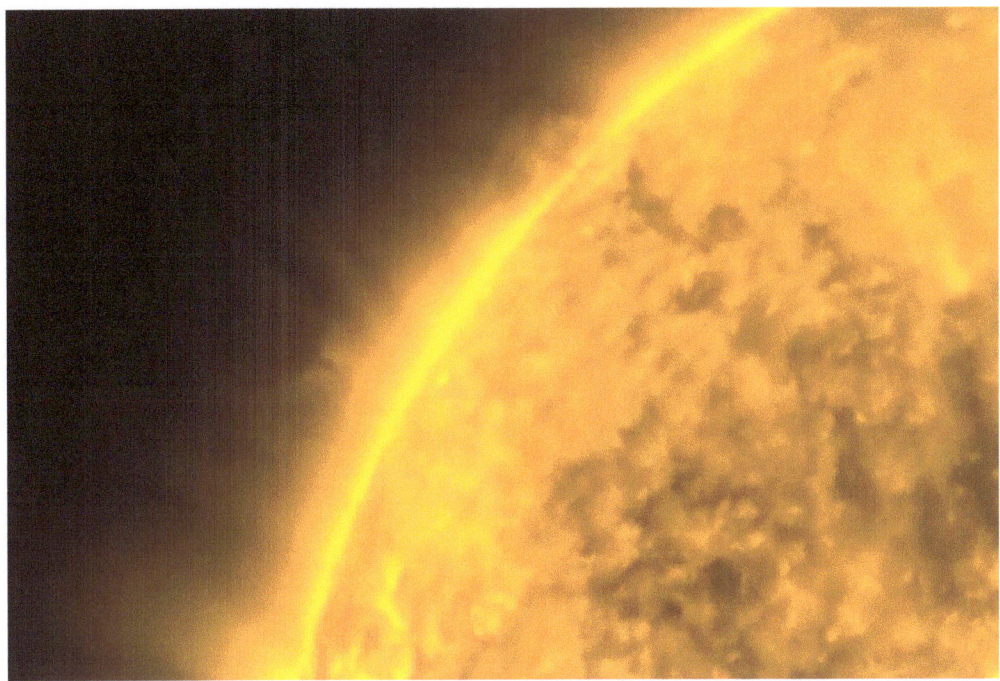

Figure 1.3: SDO image in the 171 angstrom wavelength from October 13th, 2017 showing a dark Stellar Core, which appears to be about half of the size of Jupiter.

Who can deny that these objects are in the Sun's corona? They are obviously there and anyone that says that they are not is either blind has their eyes closed or is a liar.

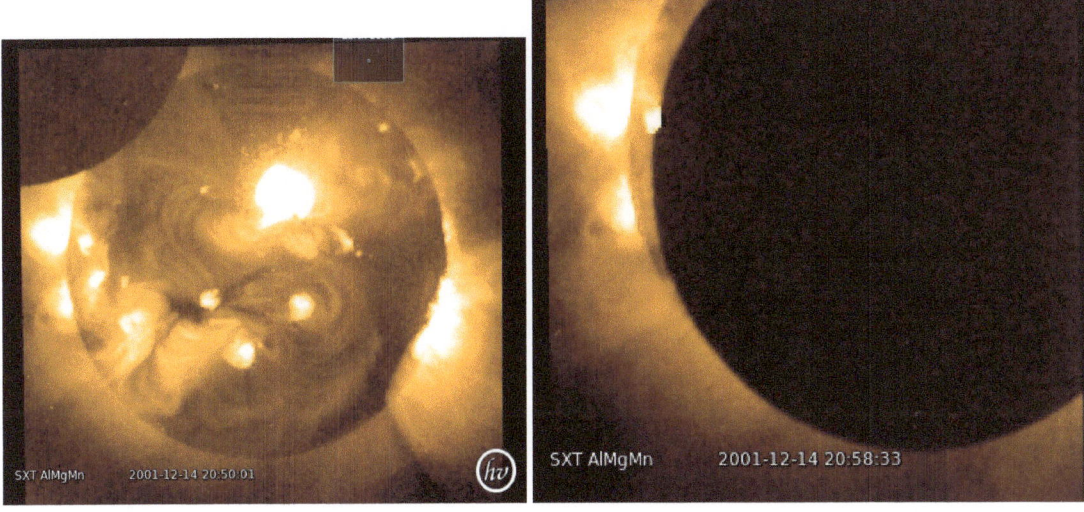

Figure 1.4. Yohkoh Satellite x-ray images from December 14th, 2001 showing an object which emits x-rays and cannot, therefore, be the moon eclipsing the Sun. X-rays are emitted from the surface, on the object's left hemisphere close to the equator, so that it forms curved contours, which indicate the curvature of the object. These seemed to also be slightly raised over the rest of the surface and indicate a layer of material covering this region of the object's surface.

But some have tried to claim that this object is the moon since a lunar eclipse was supposed to happen on this day. Now, the moon does not emit x rays, only stars are hot enough to emit x rays, but the moon does scatter a few x rays, so that in a very long exposure soft x ray image, the moon may look as shown in figure 5 below.

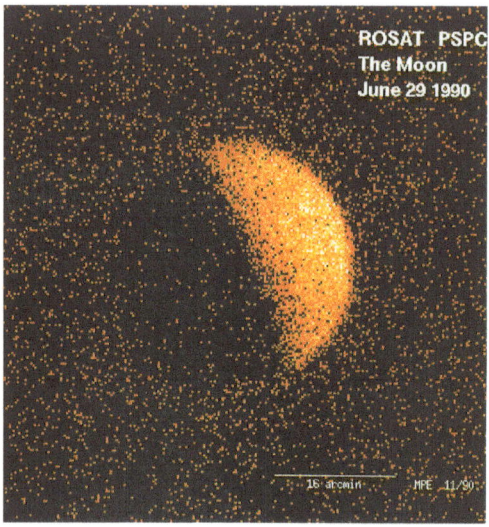

Figure 1.5. Image of the moon in soft x-rays: This is a long exposure and therefore low-resolution image of the moon, no contours or surface features of the moon can be distinguished. The number of scattered x rays from the moon's dark side is extremely low.

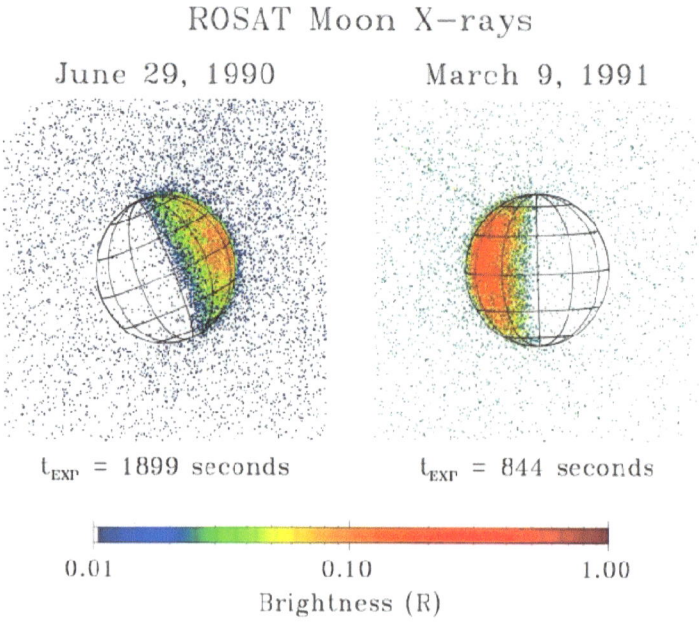

Figure 1.6. Because of the low amount of scattered x ray light, it is necessary to use a very long exposure to obtain soft x ray images of the moon. The exposure time for the image shown in figure 5, from June 29, 1990, is 1899 seconds or 31 minutes and 39 seconds.

The Yohkoh images are high-resolution images and therefore with an exposure time of the order of 1 second or less. The object in the right image, in figure 3, is almost completely eclipsing the Sun and therefore we are looking at its dark side. When we compare this image with the image of the moon we see that many more x-rays have been captured coming from the surface of the object than have been captured coming from the moon's night side, in a time interval which nearly 2000 times less than the exposure time for the moon image, which indicates that the object is emitting x-rays, and not scattering them, and cannot therefore be the moon. Another factor, which clearly shows that this object is emitting x-rays and cannot possibly be the moon, is the fact that surface contours can be seen from the captured x-rays. Scattered x-rays are scattered in random directions and therefore no resolution of surface features are possible but emitted x-rays allow surface features to be resolved once detected. Thus, the fact that the object's surface features, such as the raised layer, over the equator, can be seen, is another factor indicating that the object emits x-rays and cannot be the moon.

Now, that we have an object clearly between the Sun and the Earth that cannot be the moon and is emitting x-rays indicating that it must either be very hot, or it is highly ionized due to a high electric field, what is it? What other objects in the Solar System are between the Earth and the Sun, Venus, and Mercury? But these are tiny planets in comparisson with the Sun, which cannot eclipse the Sun in the way this object is eclipsing it. So what is it? Who among the astronomers and astrophysicists and all the NASA scientists can answer this simple question? What is this object when it cannot be the moon, cannot be Venus and cannot be Mercury? The silence is deafening.

Do they support my research? No, they do not? Do they explain what these objects are? No, they do not. Why are they silent? They are silent because they cannot deny what is obvious and they remain silent because they are afraid. They are afraid of what will happen to them, if they start telling the truth, because, at a minimum, they will lose access to telescopes and will not be able to continue their research, which was what happened to Halton Arp, due to his work on redshift, which clearly falsified the Big Bang model. They will most likley be blocked from publishing their research by the peer review system, which also happened to Halton Arp, as he explains in his book 'Seeing Red'. This is what the peer review system is all about. The peer review system is for the purpose of keeping the truth out of mainstream scientific research and there are many scientists who agree with that statement, Einstein was among them, as he simply withdrew his papers if anyone suggested that it would be seent for peer review. Think about it? How can truly novel ideas, which will take scientific thought forward, be kept out of mainstream physics? Simply make sure that nothing gets published in the 'good' established papers by letting those with old conventional ideas, the 'peers', decide what is good research and what is not good research. That cannot be left up to the readers, of course, that would lead to too much innovation, too much truth. The peers simply do not allow anything to be published that does not agree with comtempory thought and therefore innovation and truth are removed.

But what else can happen to those researchers who decide to go ahead and tell the world the truth? Well, they could be persecuted, as I was until they cannot stand it anymore, and leave their university positions. They could be persecuted and fired as James McCanney was, as he explained in his book, Planet X, Comets and Earth Changes. Then there are the more serious repercussions reserved for those physicists or astronomers, or planetary scientists, who already have a well established reputation and cannot, therefore, be silenced by removal from a university. Both Eugene Shoemaker, planetary scientist, who saw that the Shoemaker-Levy impacts meant that comets could not possibly be just icy snow balls, and Robert Harrington, astronomer, who must have found the objects that we now know are here, the Planet X objects, died, under very suspicious circumstances, which to me smells of murder covered up as a head-on collision in a very out of the way place, with no witnesses, in the case of Dr Eugene Shoemaker, and an impossibly fast progressing cancer, in the case of Dr Robert Harrington. The astrophysicists know this and so they stick to their safe research and keep silent.

In conclusion, the evidence that the Planet X System of Stellar Cores, as I have called the objects, which are clearly observed in the Sun's corona and in the inner Solar System, is here, is irrefutable. The astrophysicists and astronomers who have absolutely nothing to say about it, keep silent because they cannot refute it. And, they keep silent because they are afraid of what will happen if they do acknowledge the truth.

Dr Claudia Albers

Planet X physicist

September 21st 2018

Chapter 2

275. Planet X debris field impacting earth

The Planet X System of Stellar Cores has been generating debris in the Solar System and this debris has been entering earth's atmosphere and producing strange cloud formations, which emit light and are therefore luminescent. It seems to have started in 1850, when luminescent clouds were first observed at very high altitudes, and which have been referred to as noctilucent clouds [1]. Thus, the year of 1850 marks the likely beginning of this system starting its invasion of our Solar System (see Article 272: Noctilucent clouds and Planet X debris in the earth's atmosphere) [2]. But there is now additional evidence that Stellar Core debris is continuously reaching earth, and entering our atmosphere, which I will present in this article.

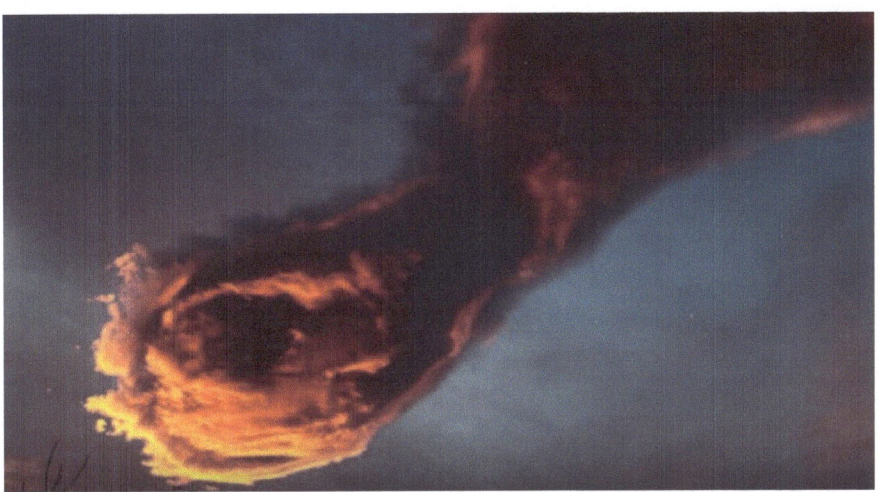

Figure 2.1. Strange cloud in the earth's atmosphere: Different portions of this cloud are emitting light in various different colors: orange, red, pink and dark blue. This cloud cannot be produced through any normal mechanism, in the earth's atmosphere, but the entrance of Stellar Core debris, which is depleted in energy, and electrons, the mechanism of water condensation around the dust, or small rocks, entering the atmosphere, together with electron capture, is able to explain the formation of these strange clouds in our atmosphere.

The water condensing around the Stellar Core debris would lose electrons to the debris, and would thus become electron depleted, as well. Then, as the water absorbs electrons, from other atoms, in the atmosphere, it would emit photons, and thus give off the light, of various wavelengths or colors. It is also possible that light in wavelengths, not detected by the human eye, would be given off by some of the debris interacting with the earth's atmosphere, such as ultraviolet light, which may then explain why more UV C radiation is reaching the surface of the earth, than can reach the top of the atmosphere, after being emitted by the Sun (see Article 264: Planet X affecting earth and radiation exposure concerns) [3].

Below are graphs showing particle density data, coming from the DSCOVR (Deep Space Climate Observatory) spacecraft, which is in the L1 position, and is, like, the ACE (Advanced Composition Explorer) spacecraft, designed to monitor incoming Solar Wind conditions. These graphs show that both on June 12th and June 16th, 2018, sharp particle density peaks were detected. This would mean that a sudden and very short lived high density stream of particles went past the spacecraft, on those two occasions. This is consistent with a small cloud of debris coming in toward Earth. It is not consistent with Solar Wind particles, as the Solar Wind density goes up and down but over a much greater time span, and would therefore not create these intense sharp peaks.

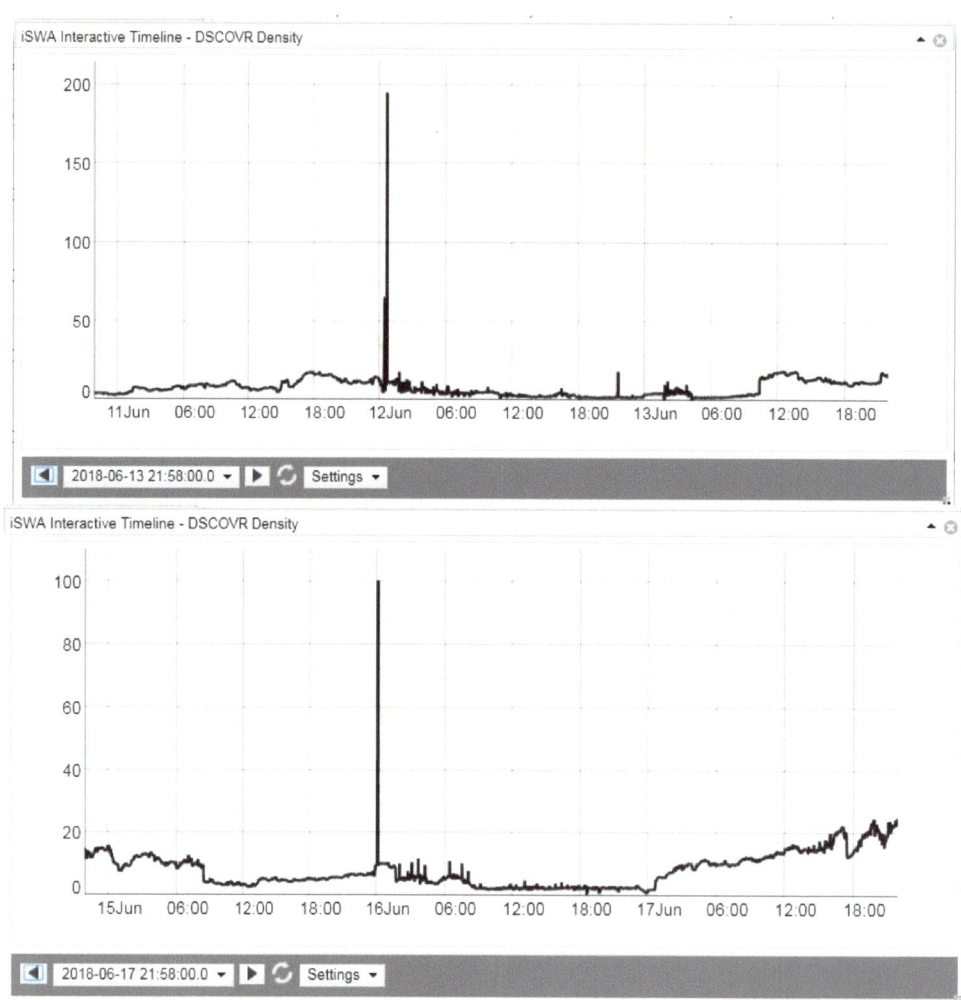

Figure 2.2. Very sharp and intense particle density peaks, in the DSCOVR data, are consistent with small clouds of debris moving past the spacecraft, on their way to the earth's atmosphere. This debris would then react with the matter, in the earth's atmosphere, thus creating the colorful clouds like the one seen in figure 1.

Since Stellar Core matter leads to the creation of clouds, which give off light, in various bright colors, which would not normally be observed, in the Earth's atmosphere, the observation of those colors, in the earth's atmosphere, thus signals the continual entrance of Stellar Core debris, into the Earth's atmosphere.

Figure 2.3. Bright strange colors observed in the earth's atmosphere, which signal insertion and absorption of Stellar Core matter, into the earth's atmosphere. Pink, orange, bright blue clouds in the atmosphere can only be explained if one understands that Stellar Core matter absorbs energy and is electron depleted and will thus cause the emission of light, in these colors, when it comes into contact with the earth's atmospheric matter. The reaction is electrical so it will also, most likely, often lead to the creation of electric storms, in the atmosphere, which will also lead to lightening in colors that would not normally be observed in the earth's atmosphere.

It should not be surprising that the Stellar Core debris is continuously reaching earth. As I have detailed in Article 259: Large Planet X Object and debris in the inner Solar System [4], the Planet X System Stellar Cores shed their outer layers of material, and have, as a result, filled the Solar System with a huge amount of debris. This debris can usually be seen in most sun observing satellite images.

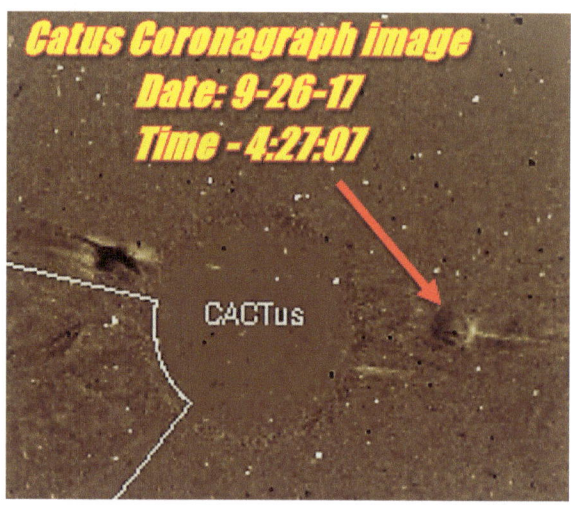

Figure 2.4. Bright spots in the image are due to the debris generated by the Stellar Cores in the Sun's corona which shed their outer layers of material. Bright spots in front of the Sun's position cannot, of course, be stars shining through the Sun.

Stellar Core matter is energy absorbing, and as this matter interacts with the Earth, it will absorb earth's energy, which it generates in its core, and will lead to the earth increasing its pace of energy generation, thus causing the Earth to increase in temperature, at all levels. This will, in turn, increase the severity of storms, increase earthquake and volcanic activity, and as the energy continues to drop, this occurrence will keep increasing both in severity and frequency. Also, as earth's energy drops, so will its gravitational influence, which will cause the earth to expand, which will, in turn, cause it to fissure and sinkholes to appear, in ever increasing frequency (see Article 247: The effect of the Planet X System on the earth) [5].

Figure 2.5. The earth breaking open, or fissuring is a consequence of its energy being absorbed by Stellar Core matter.

In conclusion, Stellar Core debris is continuously reaching the Earth and entering the atmosphere where it reacts with the matter in the atmosphere giving rise to cloud formations, which then give off the light in several bright colors, which would not normally be observed in the atmosphere. It thus becomes clear that the Earth is now being impacted, by the Planet X System debris field, which continues to fill the Solar System.

References:

[1]   http://climate.arm.ac.uk/publications/noct-paper-rev.pdf
[2]   Albers, C. (2018). Article 272: Noctilucent clouds and Planet X debris in the earth's atmosphere.
[3]   Albers, C. (2018). Article 264: Planet X affecting earth and radiation exposure concerns.
[4]   Albers, C. (2018). Article 259: Large Planet X Object and debris in the inner Solar System.
[5]   Albers, C. (2018). Article 247: The effect of the Planet X System on the earth (in Book 7: Planet X The effects on the Earth and the Sun).

Chapter 3

282. Earth in upheaval: magma rising from beneath

It was reported, at the end of 2017, that magma was rising from beneath, in the New England region, an area in the northeast of the United States, not known to be geologically active, and thus not known for volcanic activity. It is thought that the rising magma may reach the surface in the next 10 million years, which apparently is a short time in terms of geological timelines [1]. However, this may be an extremely optimistic view of what is actually happening on Earth. The African Rift Valley Region is known to be geologically active, and a region in the world where the continent is breaking into 2 parts, which will eventually be separated by an ocean. According to normal geological timelines, this is supposed to take millions of years, and yet, extremely large movement, giving rise to fissures which are appearing overnight, is occurring. This means that it may take only a few months or years for the fissures to connect and for the ocean to be able to get into these fissures, and thus cause the separation of the continent to occur (see Article 201: Africa breaking up: a preview of what is to come) [2].

Figure 3.1. Africa is breaking up: Fissures and chasms appear overnight in the African Rift Valley overnight. Thus, a process that is supposed to take millions of years is happening at an extremely accelerated rate. At this rate, it may only take a few years or months for Africa to be split into two parts separated by an ocean.

So, if according to geological timelines a process that is supposed to occur over millions of years is now happening, literally overnight, is it possible that the rising of magma under New England, which is supposed to take millions of years, can actually also happen at a vastly accelerated rate, so that a volcano suddenly appears in that region? And, if that is occurring in this region of the planet, can it also happen elsewhere?

In order to answer these questions, it is helpful to understand why fissures are appearing overnight, all over the earth. They are more likely to occur in regions where, the earth already had a tendency to move apart, such as in rift zones, but they are also appearing in regions where no such tendency had been noted before.

Figure 3.2. Fissures appearing overnight all over the world: India, Saudi-Arabia and New Zealand.

Rainfall is usually implicated in the appearance of these fissures, but how does rain suddenly erode away enough earth, to make fissures that are often several miles long? Rain is not a new phenomenon. There are many parts of the world that have experienced torrential rain for hundreds or thousands of years without this phenomenon being noticed. The other thing that is happening around the world, which has never been seen before is the sudden disappearance of ocean water from coastlines, which then reappears days, or hours, later, indicating that it was a tide.

Figure 3.3. Ocean recedes leaving boats sitting on mud, in the harbor in Punta del Este, Uruguay, on August 11th, 2017. The ocean came back but this extreme low tide had never happened before.

A tide can only be caused by a massive object being close to the earth. This then suggests that the cause is not within the earth but outside the earth. In fact, the cause is the fact that our Solar System has been invaded by objects, from outside the Solar System; these objects are dead planets and stars, and they belong to a system, which I call the Planet X System of Stellar Cores. These objects are depleted in energy and thus drain other objects of energy. The earth is being affected by these objects and its energy is being drained (see Article 243: Earth hosting at least 3 Planet X System Objects) [3].

Everything seems to come down to energy, which originates with the photon or light (see Article 191: The photon universe: it is all made out of light) [4]. Planets and stars generate energy in their cores through radioactive decay. The Stellar Cores have run out of this energy, so they are depleted, and thus, absorb energy from other objects that still have it.

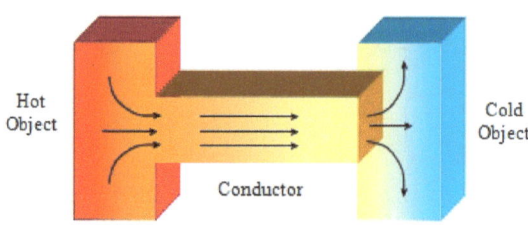

Figure 3.4. Just as heat flows from a hot to a cold object, energy flows from the Sun to a Stellar Core. Thus, more energy reaches the surface of the Sun, which causes it to become 'hotter' or more active on the surface. The draining effect causes the radioactive decay rate to increase (see Article 240: Planet X System effect on radioactive decay rate and heating of planets) [5].

This causes heat to flow from the earth's core, which then causes all layers to heat up. The first layer, after the inner core, which is the layer generating the extra heat, to heat up is the earth's liquid layer or the earth's outer core. Increased heat will lead to expansion and increased pressure, and will thus cause magma to rise, which in turn, will cause increased volcanic activity on the surface, so existing volcanoes will experience more eruptions, dormant volcanoes will become active, active volcanoes may experience more, and longer lasting, eruptions, and new volcanoes will develop, as magma rises and reaches the surface in places where it had not done so before.

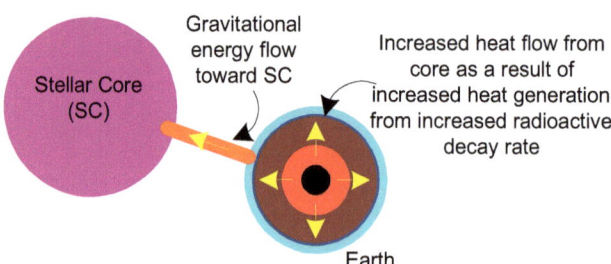

Figure 3.5. As the Stellar Core draws more and more energy, the energy flowing outwards from the core increases, and thus more energy flows through each layer of the earth, which causes each layer to experience an increase in energy, or heat, which then translates into a temperature increase (particles move or vibrate faster). This causes the magma layer to get hotter and to expand.

But this is not all. In the course of observing the Stellar Cores in the Sun's corona, it has become clear that their low energy status has greatly decreased their gravitational influence on other objects (see Article 210: Stellar Core gravity: tidal and G is not constant) [6]. This means that as the earth's energy is absorbed, its gravity will decrease, which will cause not only its ability to attract other objects to decrease but also, its ability to pull its own matter inwards to decrease. This means that internal pressure will not be countered with the same degree of inward gravitational pulling, causing the earth to expand outwards; the surface will thus be forced to grow in size, which will then cause it to the fissure, or crack open. But the expansion will be greater at the surface but will occur underground as well. Thus fissures will form at all depths. Deep underground fissures will make the movement of lava, from deep within the earth to the surface, easier, as it finds more weak areas through which to flow upwards.

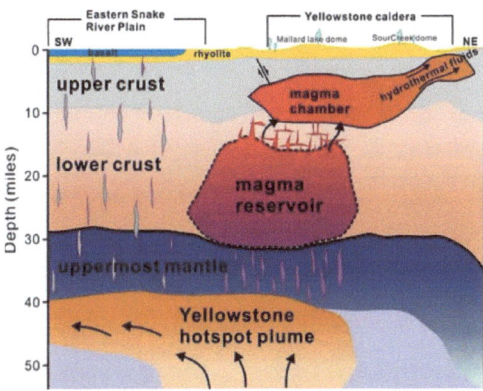

Figure 3.6. Fissures in the earth allow magma rising through the earth's mantle to fill the magma reservoir under Yellowstone, magma in the reservoir then slowly fills the magma chamber, when the pressure in the magma chamber reaches a critical level an eruption will occur.

The earth opening and breaking up in addition to the heating of the layers will also allow gases that may have been trapped underground, such as methane, to now escape to the surface; as methane mixes with oxygen, explosions become likely. This is the accepted reason why large numbers of sinkholes appeared in northern Russia, in 2015. However, there may be more than that involved in this, as in order for such large holes to appear matter filling these holes would have to have somehow disintegrated, which does not seem possible.

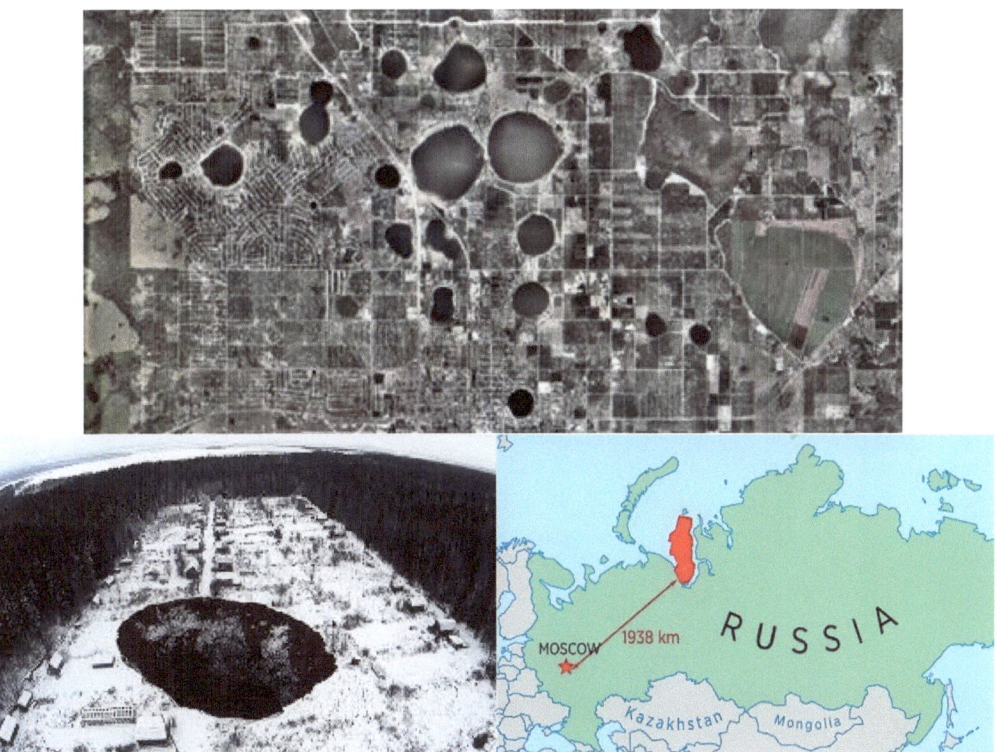

Figure 3.7. An epidemic of sinkholes appearing in northern Russia appears to have been caused by methane explosions. The heating of the earth from the core caused by the energy drain by the Stellar Cores affecting the earth will lead to gases being released.

Now, the Planet X System appears to be a huge system, with perhaps many thousands of objects, within it, and more of these objects seem to continually be arriving at the Sun's corona, so more will most likely continue to reach earth, and start interacting with it, and thus increase the energy drain, which will have the effect of increasing the heating of the earth, from the core, and cause the rate of expansion to continue to increase. Thus, as the earth continues to expand and magma inside the earth continues to heat up, more and more, magma will find its way to the surface, and eruptions may become endless events of magma, from deep within the earth, finding its way to the surface, as seems to be occurring in Hawaii.

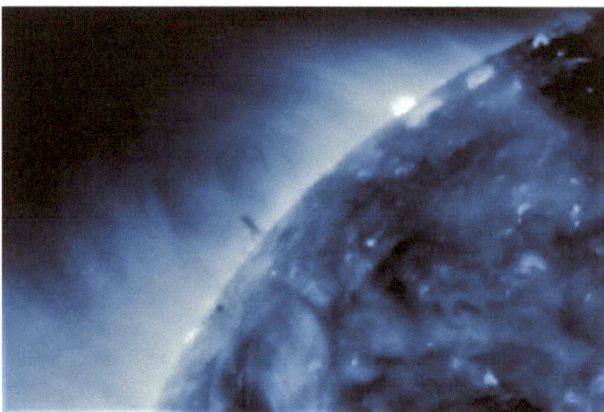

Figure 3.8. SDO image, in the 211 Angstrom wavelength, from September 27th, 2017, showing: a spherical object, a Stellar Core, making a connection with the Sun, and thus drawing energy from the Sun. These objects will do the same to the earth.

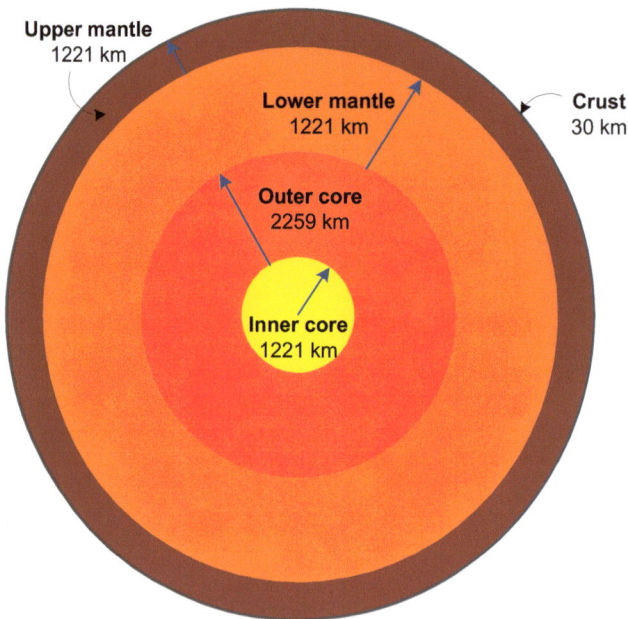

Figure 3.9. Earth's inner layers: The outer core is liquid magma. As heating increases, a larger part of the lower mantle will likely turn into liquid rock or magma. Thus, the amount of magma deep within the earth will increase.

As heat continues to flow out of the earth's inner core at an increasing rate, the lower part of the lower mantle will also become molten rock or magma, and thus, the earth will contain an increasing amount of expanding magma, which will tend to find its way toward the surface, more and more, as a result of increasing pressure. As to the increasing rainfall that is often blamed for the earth fissuring, it is also a consequence of a larger amount of heat coming from the earth's core and rising magma leading to volcanic eruptions on the ocean floor, which, in turn, heats the ocean and causes unprecedented climate change. In addition, the Stellar Cores directly interacting with the earth's atmosphere create electric effects (ionization and condensation as energy is drained out of water molecules leading to cloud formation), which lead to increased cloud formation, increased lightning and thus severe storm activity, which will drop very large amounts of rain, very quickly. In other words, the Planet X System is causing the fissuring, the increasing volcanic activity and the increased amount of rain.

In conclusion, the Planet X System of Stellar Cores has caused our planet to be in upheaval, as a result of their energy draining effect. This has the consequence of causing magma to rise from deep within the earth, causing dormant volcanoes to start erupting, eruptions to turn into unending events, and new volcanoes to form where none existed before.

References:

[1] https://news.nationalgeographic.com/2017/12/magma-bubble-rising-under-new-england-volcanoes-science/

[2] Albers, C. (2018). Article 201: Africa breaking up: a preview of what is to come (in Book 6: Planet X Physicist Articles Part 1).

[3] Albers, C. (2018). Article 243: Earth hosting at least 3 Planet X System Objects (in Book 7: Planet X The effects on the Earth and the Sun).

[4] Albers, C. (2018). Article 191: The photon universe: it is all made out of light (in Book 6: Planet X Physicist Articles Part 1).

[5] Albers, C. (2018). Article 240: Planet X System effect on radioactive decay rate and heating of planets (in Book 7: Planet X The effects on the Earth and the Sun).

[6] Albers, C. (2018). Article 210: Stellar Core gravity: tidal and G is not constant (in Book 6: Planet X Physicist Articles Part 1).

Chapter 4

291. The sun disappears: day turns into night

On July 20th, 2018, the day turned into night, in the Yakutia northern region of Russia. This region is above the Arctic Circle and, at this time of the year, is supposed to have daylight for nearly 24 hours a day. However, from about 11:00 am, it started getting dark and by 11:30, it had become completely dark; the darkness lasted until 2:00 pm. Locals talked about the possibility that an eclipse had occurred but it only affected two districts in the Yakutia region, and an eclipse should have affected a much wider area. The other possibility was that a thick cloud of dust came over the region, but officials said that there was no cloud of dust and that fires, burning in other regions, had not affected the area, and also that any pollution resulting from those fires had not reached the area. However, locals reported seeing thick layers of dust on the ground after the event.

Figure 4.1. Day turned into night, in two districts of the northern Russian region of Yakutia, above the Arctic Circle. Locals reported that the darkness had a rich yellow undertone to it

Local people also reported that the darkness was different from anything that they had seen before and that the darkness had a rich yellow undertone to it. This is most likely why the sky looks red in some of the photographs taken during the event. People also reported feeling 'a heaviness' in their chests from being outside, which made them want to go inside. This suggests that there was something in the air, which was toxic to human lungs, and the fact that locals reported seeing dust on the ground, suggests that the dust that may have been in the air, at the time, was affecting people's lungs. But people did not actually report seeing or feeling dust fall on them so it must have been very fine dust and any cloud must have been quite diffuse.

This does not seem to be a normal cloud of pollution, or dust, or smoke though, as no one reported seeing a cloud approach. It just became darker. A cloud of dust or smoke coming towards a certain region can usually be seen approaching, since this move horizontally and thus parallel to the ground, but no one reported seeing any cloud of dust, or smoke, approaching, which suggests that it came from above and therefore most likely from space. This was most likely a cloud of dust that entered the atmosphere above this region of Russia.

Figure 4.2. The two districts in the Yakutia region in northern Russia where day turned into night.

The rich yellow undertone of the darkness must have been what caused the sky to look red, in some of the photographs taken during the event, and suggests that whatever was in the atmosphere was reacting and emitting light. A cloud of normal pollution would be expected to make the sky darker and grey, not yellow or red. So, the dust was reacting and giving off light suggesting that energy was being transferred, causing electrons to move to different energy levels. Dust that comes from above and thus most likely from outside the atmosphere, and also reacts with atmospheric particles, so that radiation is emitted, is suggestive of Stellar Core dust coming into the earth's atmosphere. Stellar Core matter absorbs energy, which causes condensation, ionization and light emission. Stellar Core dust has been coming into our atmosphere, for quite a long time, and has thus been producing luminescent clouds and strange colors in the sky (see Article 275: Planet X debris field impacting earth) [1]. However, this dust usually remains suspended in the atmosphere, for a long time, as it is usually very low in gravitational energy and does not, therefore, fall to earth, like normal earth matter, until it has absorbed some gravitational energy, from interacting with earth matter. This may then be why it took 3 hours for the dust to reach the ground.

But, as mentioned before, if the dust had caused the darkness due to its density so that light from the sun was not able to penetrate it, people should have reported seeing and feeling the dust around them and falling on them, but none of the reports seem to suggest that people were immersed in a thick cloud of dust, which suggests that there was not enough dust to justify the darkness that ensued, so what could have obscured the sun so that day turned into darkness? What most likely happened, then, is that the local sun simulation was interrupted by the incoming dust, in other words, the sun simulator malfunctioned as a result of the incoming Stellar Core dust, most likely because the dust absorbs energy, and thus draws a current of electrons from the environment. The main simulator, in operation, is the one in orbit. This device seems to transmit power and sometimes project a laser beam onto clouds, thus producing a holographic projection of the sun in the clouds.

Figure 4.3. Sun Simulator in orbit: the reflection it creates on the earth is at least 21 times larger than the reflection the real Sun would produce, from the satellites perspective. The simulator's reflection has an estimated diameter of 791 mi (1274 km), whilst the real Sun's reflection would have a maximum diameter of 37 mi (59 km), showing that the reflection on the atmosphere cannot be of the real Sun (see Article 224: Sun simulator in orbit: irrefutable evidence) [2].

Figure 4.4. The Sun is both behind and in front of clouds, which is only possible, if this Sun is not the real Sun. It is actually a holographic projection of the real Sun.

The fact that it is a holographic projection can be seen from the sequence of photographs below in which an airplane flies right through this type of sun simulation, in the sky.

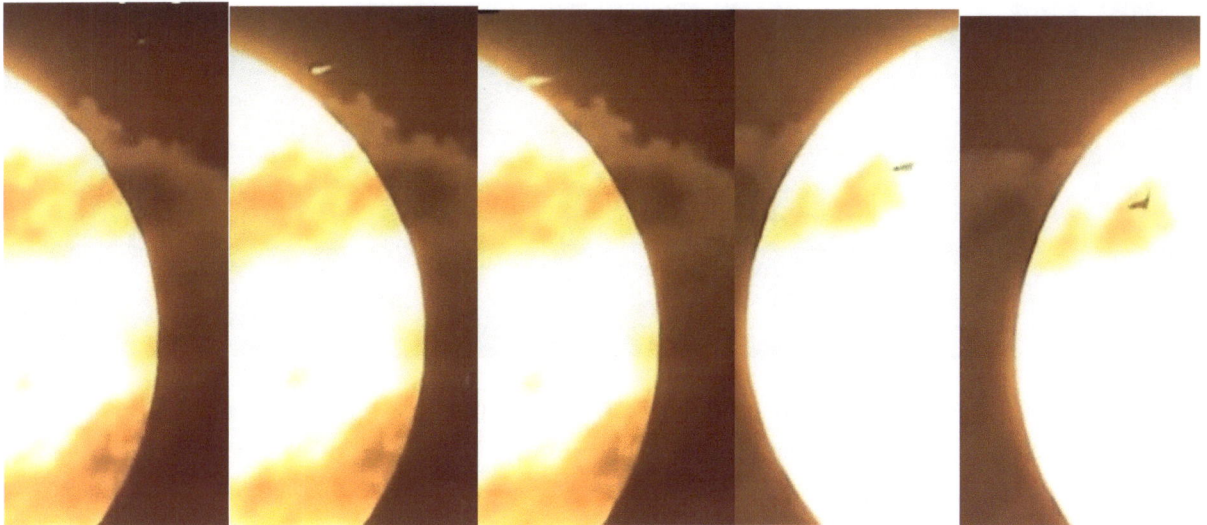

Figure 4.5. An airplane, which seems enveloped in a cloud, approaches the Sun simulator. The airplane seems to sharply increase in size, as it approaches, indicating the presence of a magnifying lens, between the Sun simulator and the camera. Once inside, the airplane is not seen, for a moment, and then begins to emerge from the simulator, but from a point closer to its left side (see Article 233: The holographic Sun in the clouds) [3].

Now, if dust falls through the atmosphere, and interrupts the projection or signal from the simulator in orbit to the clouds below, it is to be expected that this kind of holographic projection in the clouds would be interrupted resulting in the holographic sun going out. However, the amount of dust that seems to have come in does not seem to be significant enough to cause the day to turn into night and thus the darkness that ensued must have occurred only because the sun simulator, illuminating these two regions, failed. Thus, the extreme darkness produced by this incoming dust, suggests that the real sun is now extremely dim, or it was not emitting any light at all, at the time, and therefore the only illumination the atmosphere was getting was from the simulator. In which case, it would be likely that only the lower portion of the atmosphere would have been illuminated, as the simulator would be at cloud altitude and the atmosphere extends very far above that. But, in order for the sky to look blue from the ground, only the lower part of the atmosphere would need to be illuminated.

Now, the fact that the Sun goes completely dark at times became obvious to me, when I first noticed that the, so called, 'SDO eclipse season' was actually the Sun going dark. The fact that the corona shrinks back as darkness progresses across the sun, indicates that it is not an eclipse that is affecting the Sun, but the Sun is actually going dark.

Figure 4.6. SDO images of the Sun on August 16th, 2017, the first day of the eclipse, as darkness progresses across the Sun, the corona recedes showing that light emission decreases in the region on the Sun which is about to go dark, first.

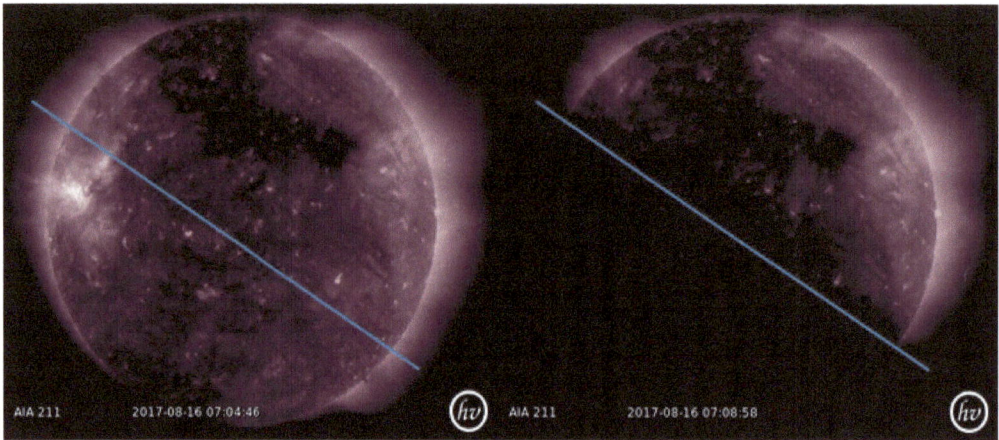

Figure 4.7. SDO images 4 minutes apart showing that the Sun's corona shrinks back, instead of being covered by the earth, as we would expect from an eclipse. The earth, which is supposed to be producing the eclipse, cannot cover the region of the corona, on both sides of the Sun, which is missing. Complete cell like structures is seen at the edge of the darkness, suggesting that these structures turn off sequentially, as the Sun becomes progressively darker.

The SDO eclipse has been occurring since 2011 and therefore the Sun must have been turning dark, at least, since then. It turns dark for up to an hour a day during the 'eclipse season'. However, it is likely that it also turns dark at other times but that the system of simulators, illuminating our atmosphere, does not allow us to see that the Sun is not emitting light. In this case, the Sun must have remained dark for at least 3 hours. Since the Stellar Cores will continue to draw energy from the Sun, and thus continue to weaken it, the Sun will most likely stay dark for longer and longer periods of time, until it will eventually remain dark most of the time and ultimately go out completely and for good. At the moment, it is also likely that when the Sun does emit light, that this light is very dim (see Article 195: Stellar Cores and the dying Sun) [4].

The fact that this event occurred suggests that greater quantities of dust have started coming into the earth's atmosphere than has happened before, and it is therefore likely that the amounts of dust coming in will thus increase and may also become more frequent, in which case, failure of the sun simulator system will occur more often and more of the earth's population will see day suddenly turn into night.

In conclusion, an event in which day turned into night, in a northern part of Russia, seems to be due to Stellar Core dust entering the atmosphere, which then led to sun simulation failure. The incident also shows that the Sun is either extremely dim or going completely dark for extended periods of time. Stellar Core dust entering the atmosphere is likely to keep increasing, which will most likely lead to this type of event repeating itself with increasing frequency.

References:

[1]   Albers, C. (2018). Article 275: Planet X debris field impacting earth.
[2]   Albers, C. (2018). Article 224: Sun simulator in orbit: irrefutable evidence (in Book 6: Planet X Physicist Articles Part 1).
[3]   Albers, C. (2018). Article 233: The holographic Sun in the clouds (in Book 7: Planet X The effects on the Earth and the Sun).
[4]   Albers, C. (2018). Article 195: Stellar Cores and the dying Sun (in Book 6: Planet X Physicist Articles Part 1).

Chapter 5

304. Can Planet X cause asteroid type impacts?

The Planet X System Objects or Stellar Cores are able to hover close to the Sun's surface inside the inner corona, for long periods of time and draw matter from the Sun, through a connection which looks like a root. This type of connection is also in the shape of a vortex, which indicates a weak gravitational attraction between the object and the Sun, and the fact that they are so close to the Sun when the connection is observed indicates that the attractive force is tidal in nature. But the lack of collisions indicates that they are, at the same time, also repelled by the Sun. The repelling force appears to be electric in nature and due to the fact that both the Sun and the objects have positively charged surfaces, and since the electric interaction causes two positive charges to repel, just like 2 North Poles of a magnet repel, the two objects are repelled away from each other.

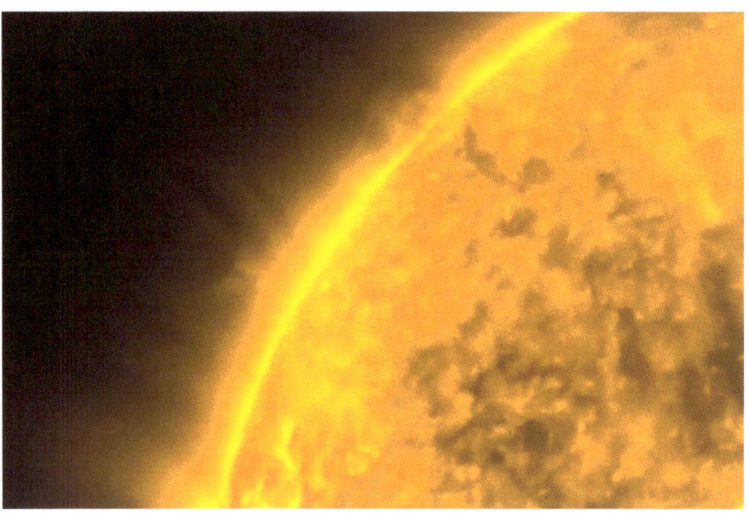

Figure 5.1: SDO image in the 171 angstrom wavelength from October 13th, 2017 showing a dark Stellar Core, which appears to be about half of the size of Jupiter. The makes a vortex or root like plasma connection with the Sun and hovers close to the Sun without impacting it showing that there must be a force repelling it away from the Sun.

This repelling effect between celestial objects is likely to extend to all objects that have a core in them which is still generating energy through radioactive decay. As the gamma ray photons are released by the nuclei splitting into two, electrons belonging to neighboring atoms, absorb the photons. This causes the electrons to leave the atom, they were in, and move outwards toward the outer layers of the object. This leaves behind atoms with missing electrons, and thus positively charged so that a positively charged surface is created. Objects that do not have a core, in which radioactive decay is taking place, will not be able to generate a positively charged surface and an outer negative layer, and will thus impact planets. Small asteroids and comets are not likely to have enough radioactive compounds in them to generate enough energy for such a process and will thus not be repelled by planets in this way and will, therefore, be able to impact the surface. Stellar Core debris is therefore capable of impacting planets once it has gained enough energy to respond to gravity. This process, however, takes a long time, and the larger the debris piece the longer this process will take.

Figure 5.2. Asteroid coming in for impact: It is not repelled by the planet's surface because it has no core generating energy through radioactive decay that will create a positively charged surface.

Stellar Cores seem to be extremely depleted in radioactive compounds to decay, and therefore are not able to generate energy in their core, but since they have over their lifetime been able to create a positively charged surface, as a result of this process, at the end of their lives they will still have a positively charged core, which is extremely deficient in electrons. So, when they reach the Sun's corona, they will powerfully draw in the Sun's electrons. This withdrawal is likely to be so rapid that the Sun's driving electric potential drops to the point that its atmosphere drops out of plasma arc mode and stops emitting light. This is what is observed during the so called SDO eclipse season (see Article 232: The Sun can go dark: the implications) [1]. The electrons gained by the Stellar Core, finding themselves, in a much lower energy environment will release large amounts of energy, in the form of photons, which other particles making up the object will absorb. As the electrons lose gravitational energy they will move inwards toward the object's core and be absorbed by atoms inside the core, so the object's initially gained an outer layer of electrons disappears, and it goes back to operating as a superion.

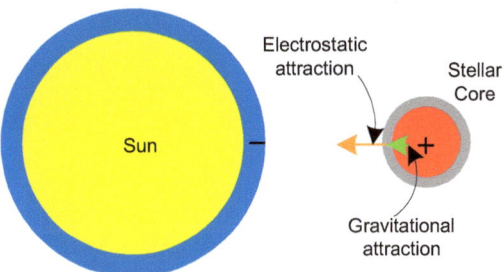

Figure 5.3. Stellar Cores are like superions and are thus electrostatically attracted to the Sun's outer negative layer. It is thus the electric interaction that causes them to be attracted to the Sun, not gravity.

In the meantime, the object starts to draw matter from the Sun through the weak gravitational attraction it is able to exert on the Sun's surface material and because it is so energy depleted, energy is drawn from the Sun to the point where the contact is made between the two. This material becomes very hot and the local electric field becomes extremely intense, which may then give rise to solar flare or CME event at that spot, which causes matter emerging from the emitted photons to be repelled by the

Sun's corona. The Stellar Core will then be repelled as well, and it will move out into space with the CME material. It will eventually be attracted back to the Sun but in the meantime, it is moving out through the Solar System taking a lot of its shed debris with it. The large Stellar Cores will return to the Sun. The smaller Stellar Cores will most likely be ejected further out and may reach some of the planets and may then interact with them and absorb energy from them. They will most likely eventually still end up returning to the Sun, as the Sun will exert a much larger attractive force than any planet.

The debris, surrounding the Stellar Cores, is spread throughout the Solar System. This debris will eventually reach the planets and start interacting with them. The matter is energy and electron deficient just like the Stellar Cores, from which they came, and will thus start drawing matter and energy from the planets. This will cause the planets' atmospheres to heat up and their gravity to start dropping which will, in turn, cause expansion and the break-up of their surfaces. This is what is occurring on earth as Stellar Core matter in the form of dust and other larger sized objects enter the atmosphere. This is leading to ever increasing fissuring of the surface and ever increasing volcanic activity (see Article 282: Earth in upheaval: magma rising from beneath) [2].

Larger clumps of Stellar Core matter that are able to stay in one piece may eventually gain enough energy to start responding to gravity and may gain enough energy to melt and to self-gravitate into spherical objects. It should, however, take a very long time for this to occur and those that may have by now gone through this process are most likely small. But they will then respond to gravity and may be drawn in by a planet. They will continue to absorb energy and thus spiral in until they have reached an equilibrium energy state with the planet, at which time they may have a stable orbit around the planet.

Figure 5.4. An object which seems to be emitting red light, and is surrounded by a diffuse cloud, is seen here in a European webcam. It was caught by Jeff P in early March 2018. The object did not move across the sky but remained in the same position for an extended period of time. Stellar Cores are also often observed to stay stationary with respect to a point on the Sun with which they have made a matter connection with (see Article 227: Stellar Cores affecting the earth and possible connection to Volcanic Eruptions) [3]. This is most likely to be a Stellar Core not debris as it looks spherical. The debris will have to absorb a lot of energy before it can self-gravitate into a spherical object and will thus behave more like a normal object and thus go into a spiraling inwards orbit rather than being able to hover above a certain point on the earth's surface.

The smaller Stellar Cores, the ones that were once living moons, which are ejected by the Sun and come close to earth may be attracted to earth, via the electric interaction, they will be attracted to the earth's outer electron layer, just like they are to the Sun's corona. These objects will most likely come into the earth's atmosphere and interact with it, in much the same way that Stellar Cores interact with the Sun so that heating of the atmosphere and radiation will be emitted. Electric discharging between the earth's atmosphere and even the earth's surface and the object is also likely to occur. The earth is too small of an object to be able to have solar flare type events, but it will certainly move toward becoming more star like than before, as its atmosphere becomes hotter and more energetic.

Figure 5. Stellar Cores within the atmosphere are likely to lead to dramatic lightning storms.

From the surface of the earth, clouds emitting radiation will form around incoming dust, which is called luminescent clouds. These first appeared in 1850, in the upper atmosphere, and were given the name noctilucent clouds (see Article 275: Planet X debris field impacting earth) [4]. The dust is likely to remain suspended in the atmosphere for a very long time until it eventually absorbs enough energy to be gravitationally attracted toward the surface of the earth. Asteroids sized Stellar Core debris will most likely stay in the atmosphere or even orbit the earth inside the atmosphere, at much slower speeds than would be expected, again, because of the very low strength of the gravitational interaction between the Stellar Core matter asteroid and the earth. They are likely to also cause cloud formation and light emission, like the dust.

Figure 5.6. Noctilucent clouds occur when Stellar Core dust enters into the upper atmosphere.

However, it is likely that with time some of the debris will absorb enough energy to start behaving more like normal asteroids, especially since this system started entering the Solar System, some 170 years ago, and will thus impact the earth like a normal asteroid would. It is also possible that with Stellar Cores moving at high speeds through the Solar System that the orbit of asteroids within the asteroid belt becomes perturbed and that these natural solar system objects may then impact the Earth.

In conclusion, Stellar Cores ejected from the Sun may reach and interact with the Solar System planets but will most likely return to the Sun. Stellar Core dust has been entering the earth's atmosphere since 1850 and larger debris particles will most likely also enter and eventually reach the ground. Larger moon-sized debris may give rise to additional moons which will spiral inwards and eventually settle in a stable orbit around one of the planets. Larger sized Stellar Core debris pieces will take a very long time to absorb enough energy to the point that they will be able to impact a planet like a normal asteroid would. But since the System has been here for some 170 years, asteroid type impacts become more probable as time goes by.

References:

[1] Albers, C. (2018). Article 232: The Sun can go dark: the implications (in Book 6: Planet X Physicist Articles Part 1).

[2] Albers, C. (2018). Article 282: Earth in upheaval: magma rising from beneath.

[3] Albers, C. (2018). Article 227: Stellar Cores affecting the earth and possible connection to Volcanic Eruptions (in Book 6: Planet X Physicist Articles Part 1).

[4] Albers, C. (2018). Article 275: Planet X debris field impacting earth.

Chapter 6

288. Image of the center of the Milky Way Galaxy reveals how the universe works

The Milky Way Galaxy is 100 000 light years in diameter and the Sun's position is thought to be about 28 000 light years from the center. A light year is the distance that light travels in one year and corresponds to 6 trillion miles (10 trillion kilometers). The center of the galaxy is much brighter than the rest of the galaxy and bulges out from the plane of the galaxy.

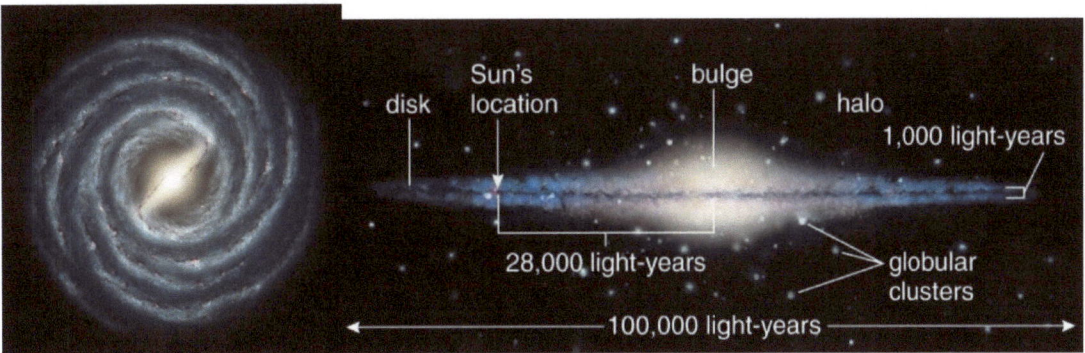

Figure 6.1. The Milky Way Galaxy is 100 000 light years in diameter. The center bulge is a region where stars and star clusters are much closer together than elsewhere in the galaxy, which causes the region to be much brighter. Globular clusters are spherical collections of stars which orbit a galactic nucleus.

Figure 6.2. A Globular Cluster: it is a compact spherical collection of stars tightly bound by gravity. Star clusters may have from 50 and 100 000 stars in it. The fact that they are tightly bound suggests that they are young structures and made of young stars as stars lose gravitational influence as they age and would expand and not be as compact (see Article 283: Planet X, planets, stars and black holes) [1].

The galactic bulge seems to extend for some 20 000 light years, but the nucleus of the galaxy is much smaller. The galactic nucleus of the brightest galaxies only extends for about 100 au, which is the size of the Solar System. The fact that these galaxies are so bright would indicate that they produce extremely strong electric fields and are therefore high energy bodies and since energy is associated with gravity they would, therefore, have stronger gravity. The stronger gravity would allow the nucleus to pull in the matter more strongly and would cause the galactic nucleus to be smaller than galactic nuclei that are not as high in energy and consequently not as bright. Since our galaxy is not one of these very bright galaxies, the galactic nucleus may be larger, perhaps 1000 au in size but this is still much smaller than 1 light year since 1 light year is 63 000 au. Therefore, the galactic nucleus is a tiny portion of the bulge. The brightness of the bulge comes from stars and clusters of stars that are very close together, much closer than they are further out from the nucleus.

Although it is widely believed that there is a supermassive black hole, at the center of most galaxies, the Stellar Cores, which have invaded our Solar System have shown that this is impossible; stars have dense solid cores, like planets do, and do not collapse gravitationally, in fact, quite the opposite happens, they lose gravity and expand, as they get older, because an object's gravity or gravitational influence depends on the photon energy in the particles, making it up, and this energy decreases as an object ages. We would, therefore, expect the youngest objects to be smaller and brighter than older objects. Thus, Active Galactic Nuclei are most likely nuclei of young galaxies.

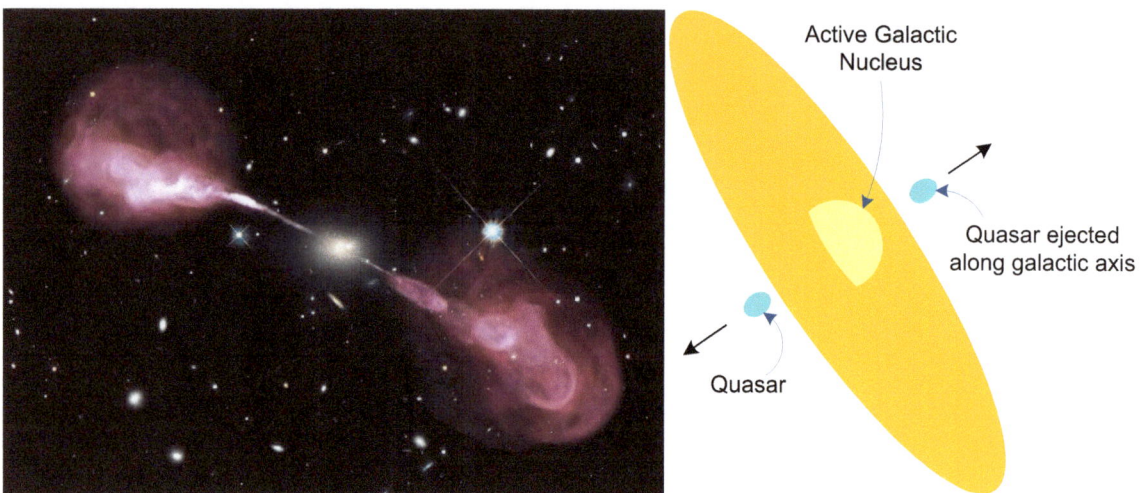

Figure 6.3. An active galaxy ejects material, which emits radio waves along the minor axis of the galaxy, from its nucleus, which is small (100 au) and extremely bright, and thus, extremely energetic. It also ejects compact globules of material which condenses into quasars, which eventually become new galaxies, as discovered by the astronomer Halton Arp [2].

If galaxies eject material that turns into new galaxies, then it is to be expected that they also eject smaller amounts of material, which turns into star clusters, and also even smaller amounts of material which turn into star systems, with stars and planets in them. It is also likely that the older galaxies, which are not as bright, or as energetic, eject smaller amounts of material, which condenses into star clusters, whilst younger galaxies eject large amounts of material which turns into new galaxies. Since our galaxy does not seem to have an extremely bright center, it is most likely an older galaxy and would be expected to eject smaller amounts of material at a time. A large number of star clusters observed close to our galaxy and orbiting around it seems to indicate that. These are most likely the product of ejections from the center of our galaxy.

Recently, a radio telescope image of the center of our galaxy, the Milky Way Galaxy, has been obtained. The image covers a region, which extends 1000 light years horizontally and 500 light years vertically. Since the actual nucleus will occupy a very tiny part of the entire region seen, it is likely that what we do not see the nucleus, but the material that the nucleus has recently ejected, and which is still close to it.

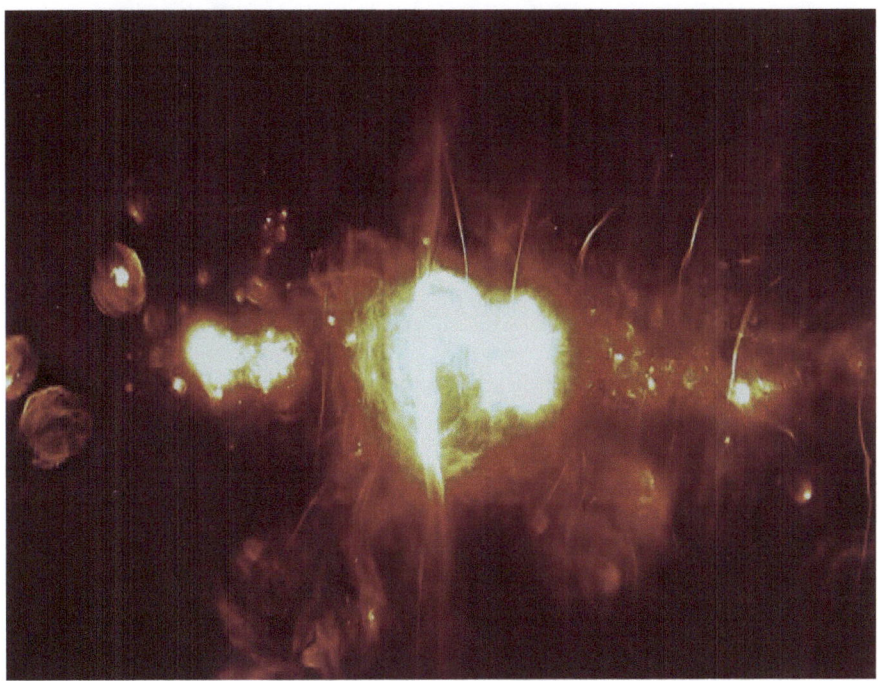

Figure 6.4. Radio telescope image of the center of the galaxy covering a region, which is 1000 light years in length, and 500 light years in height: The nucleus will most likely be smaller than 1 light year, which would make it smaller than 1 thousandth of the region seen in this photograph. We see many circular structures, which are most likely globular star clusters, in a formation stage. Long filaments can also be seen. These are most likely long ejections of plasma from condensed material not far from the galactic nucleus. This suggests that there may be more than one very large and dense galactic nuclei type structure within the center of the Milky Way Galaxy. But there is one region that is brighter and this is where the original galactic nucleus is.

In order to start ejecting light, which turns into matter, an object needs to undergo condensation so that a core is formed; plasma ejections coming from a galactic nucleus would be expected to be far out of this region before they are able to do their own ejections. But we are seeing ejections coming from different points in this region. This could be because as a galactic nucleus age, it expands, and then breaks into separate different pieces.

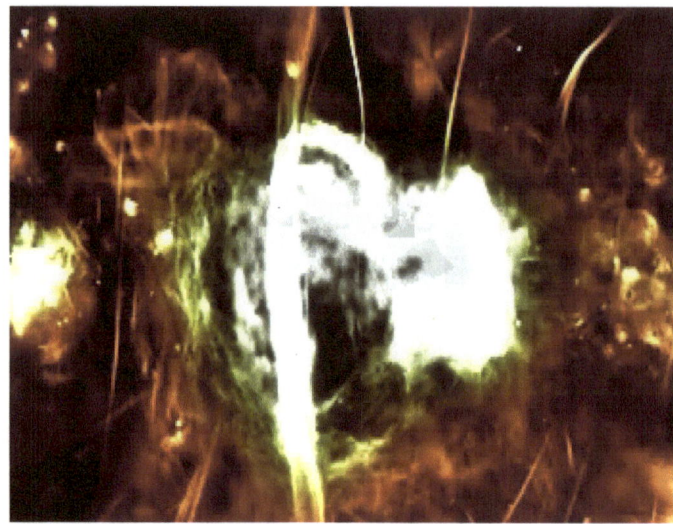

Figure 6.5. The galactic nucleus is most likely somewhere inside the brightest part of the plasma seen here. The long curving ejection is most likely coming from the most energetic region, which would be the nucleus of the galaxy. The nucleus is thus most likely situated where the long curving ejection seems to originate, in the image, or at about the center of the brightest structure in the image.

As pointed out before, the fact that galaxies have so many globular structures surrounding them suggests that these form close to the center of galaxies, and are ejected outwards, which would explain why we see so many large circular light emitting agglomerations of plasma. Many of these large structures (several light years in length) are still inside more diffuse clouds of plasma. These are most likely younger and thus closer to the nucleus, but will in time move further away from it, and at the same time condense into a cluster of stars. It is also likely that most of the ejection is along what becomes the major axis of the galaxies, or along the plane of the galaxy, and that therefore the whole galaxy is made of star clusters, that separate, in time, and populate the arms of the galaxy. Globular clusters would also tend to start out compact but with time energy and thus gravity would decrease and therefore the stars would move apart and the cluster would be much less compact.

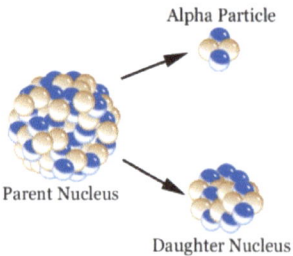

Figure 6.6. Radioactive decay is the breaking up of a heavy nucleus into 2 lighter nuclei. A drop in the gravitational energy of the particles in the nucleus will decrease the attraction between protons and allow the electrostatic repulsion between them to pull the nucleus apart.

Now, all objects in the universe seem to operate in a very similar manner, so that both stars and planets have solid dense cores and generate energy through fission or through the decay of unstable nuclei (see Article 240: Planet X System effect on radioactive decay rate and heating of planets) [3]. The gravitational interaction becomes the strong interaction, within the nucleus of an atom, and thus when the energy within particles drops and gravity drops so does the strength of the strong force, which in turn, causes the splitting of atoms or fission to occur more easily and thus more often. In other words, the radioactive decay rate increases.

Since all matter seems to operate in the same way everywhere, so that planets and stars operate in a similar manner to atoms, it is likely that a galactic nucleus would operate in the same way, as well. However, a galactic nucleus would be expected to have a much denser core, than a normal star, and thus will most likely have extremely heavy nuclei in its core. In other words, there would probably be nuclei in it that we have never encountered and with atomic numbers much higher than that of uranium, which has atomic number 92, and even higher than the highest atomic number on the periodic table, which is 118. The atomic number of an atom corresponds to the number of protons in that nucleus. Fusion is expected to occur after the light is emitted by the galactic nucleus, which then turns into matter. When the matter emerges from within extremely high energy photons, and the electric field in the environment is extremely high, an extremely strong gravitational interaction results, which leads to an extremely strong attractive force between protons, which will thus fuse into extremely heavy nuclei. As energy is expended and time goes by, lighter elements down to hydrogen will form. But the heaviest elements will condense first and thus be very abundant in the first solid matter formed. These become the cores of celestial objects.

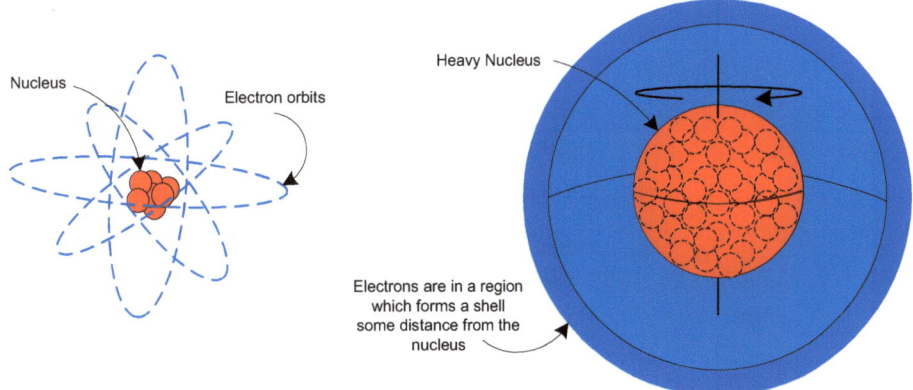

Figure 6.7. Light nuclei may have electrons in different orbital levels, and in different planes of rotation. In the case of an atom with a very heavy nucleus, the electrons will be in a region, which forms a shell, some distance away from the nucleus. The radius of an atom is 10 000 greater than that of a nucleus. Heavier nuclei are more likely to form first when the gravitational interaction is at its strongest after particles first emerge from within photons.

As matter gathers around the largest and densest cores, these turn into living stars. Smaller cores turn into planets, which operate in the same way but do not generate enough energy to emit light from their atmospheres (see Article 185: Stellar formation: Stars are formed from light and Article 186: Matter condensation: as it is with the atom so it is with a star) [4, 5].

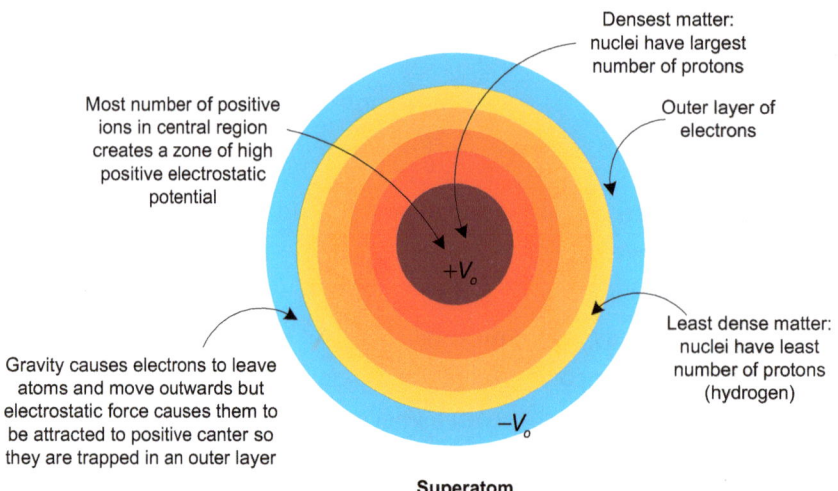

Figure 6.8. All celestial objects will have a very dense core, and less dense layers, as we move outwards from the center, in the form of concentric shells. The last layer will be gaseous and form the atmosphere of the object. The largest celestial objects will have the largest and densest cores with a greater abundance of extremely heavy nuclei (atomic numbers much higher than anything we have encountered on earth) in the case of the largest objects. The largest will also generate a higher electric field in the outer layers, and thus atmosphere, allowing it to emit extreme amounts of light, from its atmosphere.

In conclusion, a radio telescope image of the center of the Milky Way galaxy indicates that the nucleus of the galaxy is ejecting material, which forms clouds of plasma around the nucleus. Circular formations are likely to be globular clusters, in a formation stage, which will eventually end up orbiting outside the galaxy or join the previously ejected stars that have populated the plane of the galaxy. Ejections are occurring from several different points suggesting that the original galactic nucleus has expanded and broken into pieces. This agrees with the idea that gravity is associated to energy and that as an object age, it loses energy and therefore gravity, and thus expands, which then leads to objects like galactic nuclei, breaking into pieces. But everything will most likely fission so that a condensing blob of plasma ejected by a galactic nucleus eventually forms into a huge number of stars, and individual stars may fission into more than one star thus explaining why most star systems are binary or trinary.

References:

[1]   Albers, C. (2018). Article 283: Planet X, planets, stars and black holes.
[2]   Arp, Halton (1998). *Seeing Red*. Apeiron, Montreal.
[3]   Albers, C. (2018). Article 240: Planet X System effect on radioactive decay rate and heating of planets ((in Book 7: Planet X The effects on the Earth and the Sun).
[4]   Albers, C. (2018). Article 185: Stellar formation: Stars are formed from light.
[5]   Albers, C. (2018). Article 186: Matter condensation: as it is with the atom so it is with a star (in Book 3: Planet X revealed Gravity and Light).

Chapter 7

316. A rogue planet with a magnetic field 200 times stronger than Jupiter's

In 2006, an object with designation SIMP J013656.5+093347 was detected through a near-infrared/optical (visible light) survey. The object is 20 light years from Earth. From its spectrum, it was determined that the object was a T2 to T2.5 Brown Dwarf and the brightest T class Brown Dwarf in the northern hemisphere and therefore a good candidate for further study [1]. Now, Brown Dwarfs are believed to be sub-stellar objects or objects between a star and a gas giant planet. There are several different types of Brown Dwarfs: M, L, T, and Y. Brown Dwarfs emit mainly infrared light and are thus not bright in visible light. The brightest and youngest Brown Dwarfs are the M type Brown Dwarfs and the faintest and oldest are the Y type. Brown Dwarfs are thought to range in mass from 13 times the mass of Jupiter to about 80 times the mass of Jupiter but have a radius, which is very close to Jupiter's radius which would make them much denser than Jupiter.

Figure 7.1. Artist's impression of a Brown Dwarf: The objects are expected to look magenta or orange in visible light, although there are also reports that some may look blue.

In 2009, SIMP J013... was discovered to have a variability in the amount of light it emitted with a modulation period of about 2.4 hours which was attributed either to changes in surface features which may have been due to weather evolution in its atmosphere or differential rotation (different bands of atmosphere rotating at different speeds) [2].

In 2017, it was discovered that the object was part of a group of about 20 co-moving stars called the Carina Near stellar co-moving group. But, since the stars, in this group, are thought to be very young stars, i.e. no older 200 million years, the Brown Dwarf, would not have had time to burn through all its initial deuterium, to now be a T class Brown Dwarf. This meant that it had to have a much lower mass to start with. It was therefore determined that the object had a mass of only 12.7 times the mass of Jupiter and was therefore right at the threshold of the mass at which an object is supposed to be able to burn deuterium. That meant that the object may be a planet and not a Brown Dwarf. But if it was a planet, it was a planet that is not orbiting any star and was, therefore, a rogue planet [3].

In 2018, the object was studied by the VLA (Very Large Array), radio telescope, in Socorro, New Mexico, and it was discovered that it had significant radio flaring, which indicates the object's ability to produce aurora. Its magnetic field was also measured to be 200 times stronger than Jupiter's [4]. A planet that is able to produce such a strong field was not likely, as only a star is supposed to be able to be able to have such a strong magnetic field, especially if it does not orbit a star, since a star's solar wind is supposed to make a huge contribution to a planet's magnetic field.

Now, there are several problems with all of this. First of all, the mass of an object can only be measured if there is another object rotating around it, then, from the period of rotation, Kepler's third law can be used to determine its mass. But the mass attributed to this object does not come from any such process, it comes from the accepted theory of stellar evolution, which is based on stars and Brown Dwarfs being made of gas and gravitational collapse being stopped only because of thermonuclear reactions, pushing the gas outwards. This theory has been completely falsified by the presence of the Planet X System Stellar Cores in the Solar System, some of which are much larger than the Sun and the fact that they have solid surfaces, which even after having undergone expansion, still appear to be very dense. And, the fact that these objects shed any material clinging to this hard surface indicates that what is left of the original object is the densest and deepest part of the object, and therefore the core. This means that stars have solid dense cores inside them, just like planets.

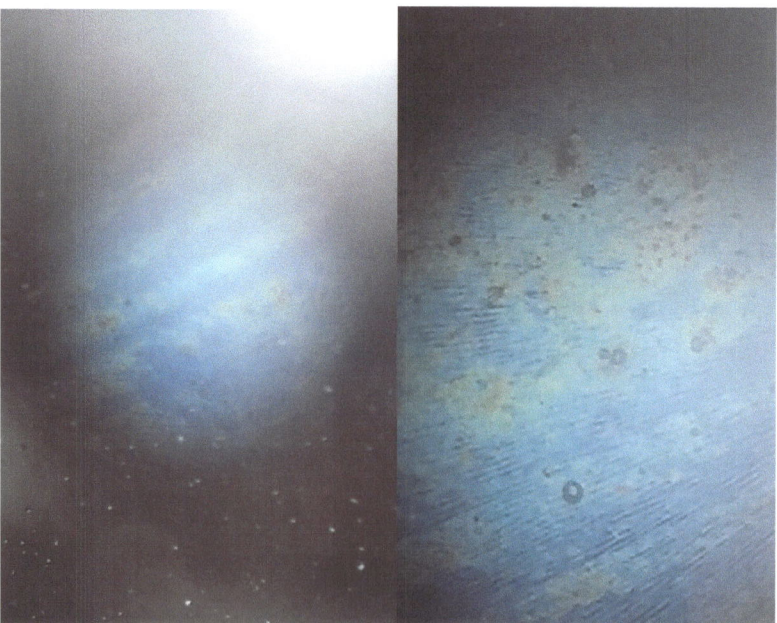

Figure 7.2. The Blue Brown Dwarf photographed through a telescope was estimated to be 3.3 times the size of Jupiter, to be shedding its top layer of material and to have a solid surface covered in fissures, which indicate that it has undergone expansion. The surface material appears to be much denser than the material in the upper layer, which is being shed and is in the form of stripes, which indicates that it is the surface of the core of the original object, a star or a planet.

In addition, the so called 'SDO eclipse season', which is not at all an eclipse, but rather, the Sun going dark, shows that there are no thermonuclear reactions occurring inside the Sun, as if they were, there would be light continuously coming from inside the Sun, and it would be impossible for its surface and atmosphere to stop emitting light.

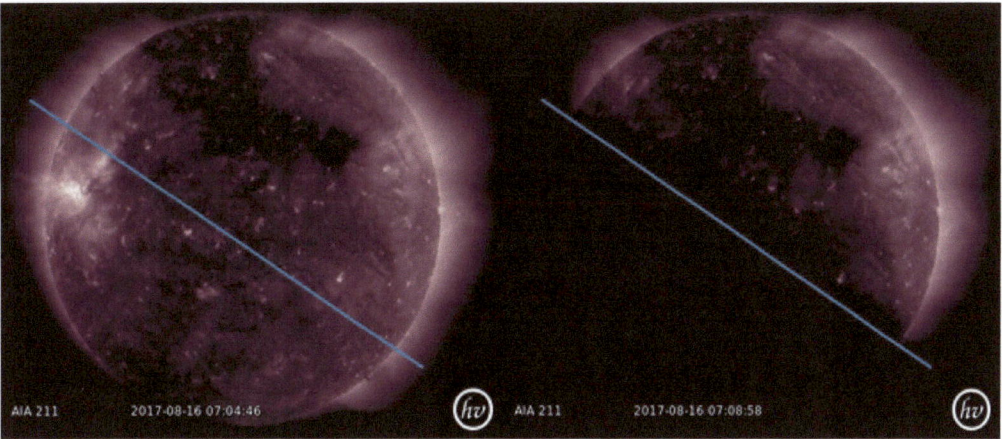

Figure 7.3. SDO images 4 minutes apart showing that the Sun's corona shrinks back, instead of being covered by the earth, as we would expect from an eclipse. The earth, which is supposed to be producing the eclipse, cannot cover the region of the corona, on both sides of the Sun, which is missing. Complete cell like structures is seen at the edge of the darkness, suggesting that these structures turn off sequentially, as the Sun becomes progressively darker (see Article 291: The Sun disappears: day turns into the night) [5].

Furthermore, interactions with Stellar Cores have periodically left gashes in the Sun's atmosphere, or corona, which reveal a semi-solid or semi-molten type material beneath the corona, which does not appear to emit the same amount of light as the corona. This means that the Sun has a gaseous atmosphere, a semi-solid layer below that, and logically solid layers below that. The core could then be several layers below the visible layers, as it is in the case of Earth. Most importantly, it means that the Sun is not a gaseous object. The Sun is more like a planet, mostly solid with a gaseous atmosphere which is emitting light due to a very large electric potential difference across it, which places it in plasma arc mode until something happens to drop the potential and the Sun then stops emitting light.

Figure 7.4. Close up from an SDO image in 171 angstroms, from August 25[th,] 2017, on the sunspot group AR2670, previously known as AR2665, with the dark region, to the left and above it, responsible for the CME of July 23[rd,] 2017. A dark circular blemish is visible close to the Sunspot group. The circular shape of the blemish suggests that it may be a crater-like structure made of a semi-solid type of material which must be made of material from a deeper layer of the Sun that is not often seen. The blemish persisted for months indicating that it could not have been discoloration of a gaseous substance (see Article 134: Planet X leaves a gash in the Sun) [6].

All this shows that planets and stars are not that different, they all have solid dense cores and any light emission comes from the atmosphere. Even the earth emits light because there is lightning in its atmosphere. Stars and planets also generate energy in their cores through the same process of radioactive decay or fission of heavy nuclei. The larger the object, the larger will be its core, and therefore the more energy it will be able to generate, which will give rise to a larger electric field in its outer layers and therefore a larger electric potential leading to increased light emission.

However, even the use of Kepler's law for determining the mass of an object will not give the correct result because observing Stellar Cores in the inner Solar System has shown that these objects do not interact gravitationally like normal objects. These objects are also very low in energy which is why they drain the Sun of energy. When a low energy object makes contact with a high mass object, energy flows from the high energy to the low energy object just like heat flows from hot to cold objects.

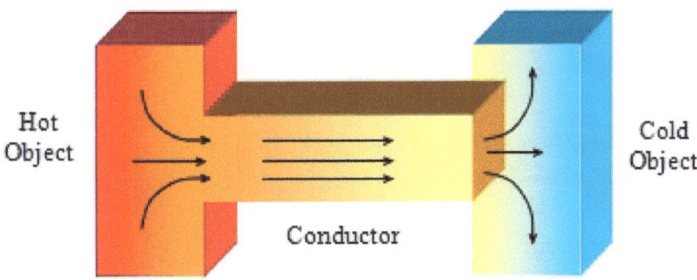

Figure 7.5. Just as heat flows from a hot to a cold object energy flows from the Sun to a Stellar Core. Thus more energy reaches the surface of the Sun which causes it to become 'hotter' or more active on the surface.

Stellar Cores absorb energy in the form of photons from the Sun, indicating that they are low in energy. They furthermore do not respond to gravity as normal Solar System objects do, which shows that the gravitational constant in Newton's Law of Universal Gravitation is not constant but is dependent on the energy, in the form of photons, inside the particles making up an object. The photons, or energy, initially come from the core and are released whenever a heavy nucleus splits into two lighter nuclei. This photon energy is absorbed by particles near to where the reaction took place and flows from particle to particle until it is distributed throughout all the matter. This sharing of energy from particle to article takes time however so that the material closer to the core maintains a higher energy and temperature than the surface, in the case of planets like Earth. This, therefore, shows that heat transfer is actually photon energy transfer, and since, photon energy is connected to the gravity, it is also gravitational energy.

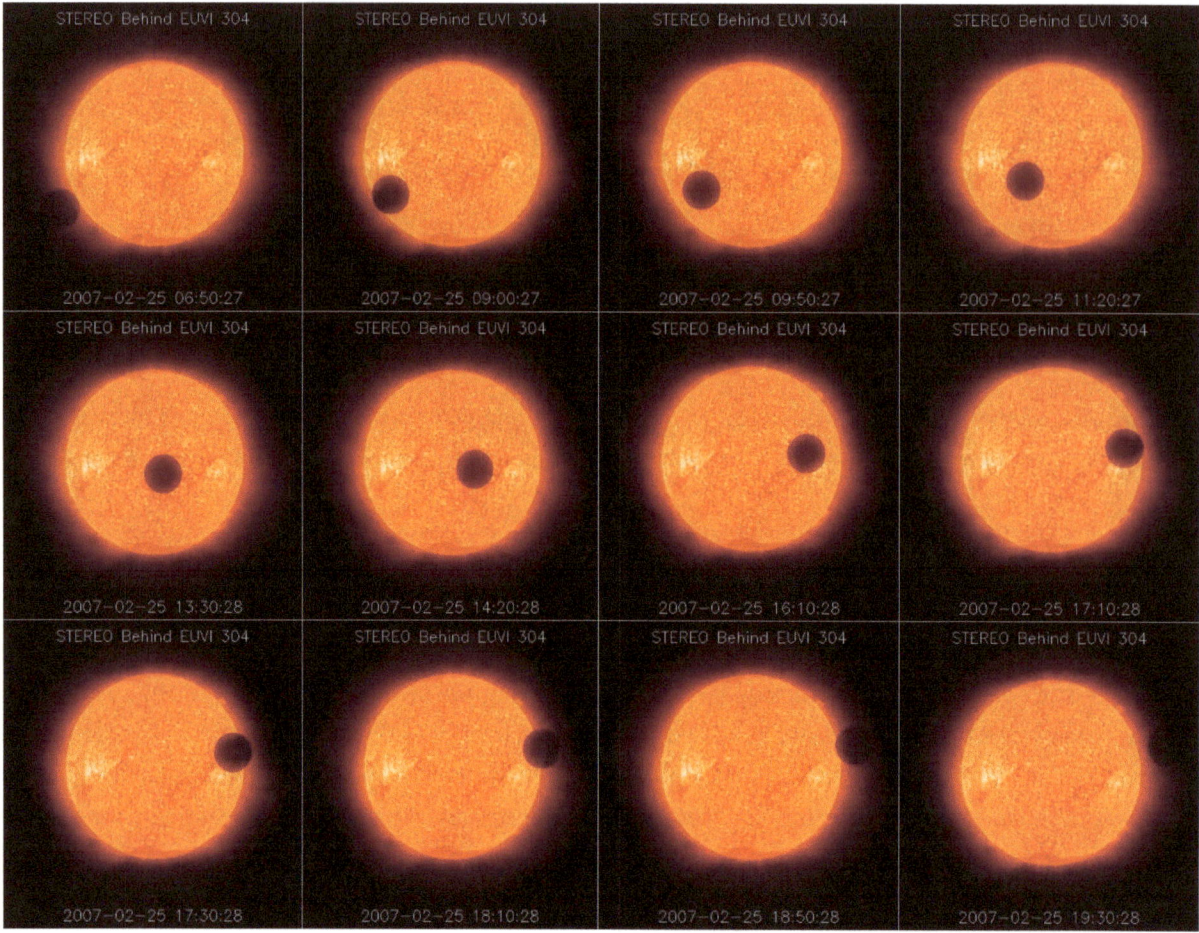

Figure 7.6. Stereo B EUVI 304 angstrom wavelength image from 2007: A large object traverses the Sun. The object is not black against the background of the solar surface, indicating that it is in the Sun's corona. The object is 2.2 times the size of Jupiter, takes 10 hours to traverse the Sun and is travelling at 39 km/s or 24 mi/s, or at a much lower speed than the Sun's escape velocity of 616 km/s. This indicates that it has a very weakened capacity to gravitationally interact. For more details, see Article 153: Planet X: Escape velocity and Gravity [7] and Article 210: Stellar Core gravity: tidal and G is not constant [8].

In the case of stars, the electric field generated in the outer layers is so high that the atmosphere becomes plasma in arc mode, which in turn generates extremely high temperatures in the atmosphere.

Thus, Stellar Cores in the Solar System have shown that the accepted theories on how stars work and evolve, are wrong, and therefore, the mass, radius, and age of stars, when determined through models based on this wrong theoretical understanding, are wrong. This means that what we really know about this object, SIMP J013... , with any confidence, since it was gained through observing the object rather than through the application of theory, is that it mainly emits infrared light and that it has a huge magnetic field. Since the energy in a young star must be higher than in an old star, and that energy will be associated with the light the star will be able to emit, from its atmosphere, a young star is likely to be bright and emit higher energy radiation, i.e. more gamma rays and x-ray, whilst an older star is likely to emit lower energy light such as infrared light. This means that this object is most likely an old star. The strong magnetic field also indicates that the object is most likely a star, as planets are not likely to have a core large enough to generate such a strong magnetic field. What this then indicates is that an ageing star's magnetic field is maintained for much longer than its electric and gravitational fields.

In conclusion, observation of the brightest T type Brown Dwarf in the northern hemisphere, when viewed through the understanding of what Stellar Core observations, within the inner Solar System, have provided, suggests that the object is an old or energy depleted star or a Stellar Core, and that stars seem to maintain a strong magnetic field for much longer than they are able to maintain their electric and gravitational fields.

References:

[1] Arttigau, E. et al. (2006). Discovery of the brightest T dwarf in the northern hemisphere. https://arxiv.org/abs/astro-ph/0609419.

[2] Arttigau, E. et al. (2009). Astrophysics > Solar and Stellar Astrophysics. Photometric Variability of the T2.5 Brown Dwarf SIMP J013656.5+093347; Evidence for Evolving Weather Patterns. https://arxiv.org/abs/0906.3514.

[3] Gagne, J. et al. (2017). SIMP J013656.5+093347 is Likely a Planetary-Mass Object in the Carina-Near Moving Group. https://arxiv.org/abs/1705.01625.

[4] https://www.sciencealert.com/nasa-discovered-a-new-free-floating-lonely-planet-with-no-host-star

[5] Albers, C. (2018). Article 291: The Sun disappears: day turns into night.

[6] Albers, C. (2018). Article 134: Planet X leaves a gash in the Sun.

[7] Albers, C. (2018). Article 153: Planet X: Escape velocity and Gravity.

[8] Albers, C. (2018). Article 210: Stellar Core gravity: tidal and G is not constant (in Book 6: Planet X Physicist Articles Part 1).

Chapter 8

317. Planet X System: Brown and Black Dwarfs with high magnetic field

In Article 315: Rogue planet with a magnetic field 200 times stronger than Jupiter's [1], I wrote about a Brown Dwarf with the designation of SIMP J013656.5+093347, which is also the brightest T type Brown Dwarf in the northern hemisphere. I concluded that the object is an energy depleted star and therefore a Stellar Core. This suggests that the large Stellar Cores we observe in the Solar System are in fact Brown and Black Dwarfs, the Brown Dwarfs would be the Stellar Cores which can only emit infrared light when they come into the Solar System and the Black Dwarfs would emit no light whatsoever.

Brown Dwarfs are known to have a strong magnetic field [2], which suggests that they are stars, not sub-stellar objects. The observation of Stellar Cores in the Solar System and their energy draining effect on the Sun has revealed that the energy inside a star's particles is associated to its electric and gravitational fields, which drop dramatically when energy is low, and since energy drops with age, so will a star's electric and gravitational fields. But the magnetic field remains strong for much longer because it is not connected to energy but to the magnetic orientation of metallic elements in the core. Since the core is solid, these would remain in the same orientation even as the photon energy in the star's particles drops. Thus, Brown Dwarfs are actually energy depleted stars and therefore Stellar Cores. This would mean that the system of objects, which I have named Planet X System, which has been invading the Solar System for at least 170 years (see Article 275: Planet X debris field impacting earth) [3], is actually a system of Brown and Black Dwarfs as well as smaller Stellar Cores, which would have been planets and moons in a star system before the central star in that system died.

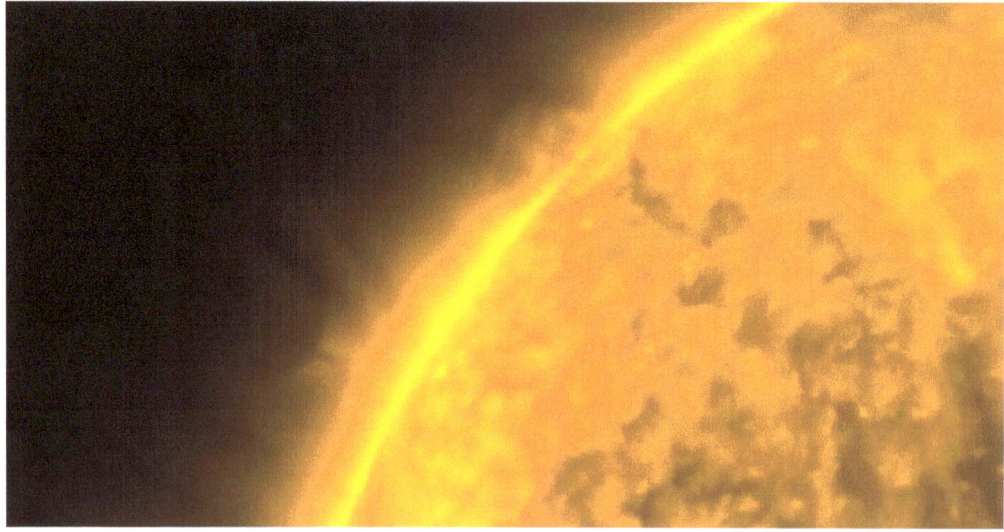

Figure 8.1: SDO image in the 171 angstrom wavelength from October 13th, 2017 showing a dark Stellar Core, which appears to be about half of the size of Jupiter.

The original theory regarding Brown Dwarfs was that they were old and ageing stars, on the way to becoming Black Dwarfs. When I studied astrophysics theory, as a student, this is what I was taught. This idea seems to have been recently scrubbed from the internet. The reason for this appears to be that the objects in the Solar System, which I have called Stellar Cores, are old stars that have lost the ability to emit light and may just be able to emit weak infrared radiation when they first arrive at the Sun's corona. These objects have also been observed, time and again, to be striped like Jupiter is, and like Brown, Dwarfs are supposed to be, as indicated in figure 8.1. Furthermore, Brown Dwarf experts believe that they would look magenta and the objects in the Solar System often emit magenta light when they interact with the earth and absorb energy from it. Is this a coincidence? No, I don't think so. What this means is that the old theory of Brown Dwarfs is the theory that more closely fits both what the Stellar Cores are and what Brown Dwarfs many light years from Earth are. They are all old or energy depleted stars and they will draw energy from living stars just like the Stellar Cores or Brown Dwarfs in the Solar System do. In other words, Stellar Cores and Brown Dwarfs are one and the same object.

Figure 8.2. A Stellar Core in the Sun's corona, about 4 times larger than earth, and thus 40% of the size of Jupiter, also has a striped appearance. The object is striped because it is an old star or a Brown Dwarf.

However, there is a way to differentiate planets from stars and that is through the ability that stars have to eject material outwards and produce a ring structure around itself. A very large and young star will then be likely to produce a large ring like structure which is called a planetary nebula. These planetary nebulae are usually associated with dead stars but the fact that a star is able to maintain such huge ring like structure indicates instead that it has a lot of energy and is most likely young. This means that supernovae may not be due to the death of a star but a birth of a star. We know what the death of a star looks like; the Stellar Cores have shown it to us. Stars end up not being able to emit light due to the loss of their electric and gravitational fields because these are associated with the photon energy, within the particles making it up. Over its lifetime a star uses this energy to generate light, a solar wind, and CMEs which feed its capacitor rings or nebular clouds but when the reservoir of unstable nuclei, in its core, is depleted the star starts dying and turns into a zombie, absorbing energy from any other living star it encounters and killing it (see Article 244: The Planet X System: destroyer of Star Systems) [4]. Now, Jupiter and Saturn seem to be small stars, as they both have small ring structures around them (see Article 132: Survey of the Orion Nebula: Jupiter and Saturn may be stars and Article 169: Planetary formation: comets to planets) [5, 6]. They both also have a huge amount of lightning in their atmospheres. The fact that Jupiter is striped whilst Saturn is not may then be an indication that Jupiter is older than Saturn and is at a stage in its life when it has already developed that particular Brown Dwarf feature: stripes.

Now, if Jupiter and Saturn are both stars, how can we say that Brown Dwarfs larger than Jupiter are sub-stellar objects? It is not possible; Brown Dwarfs have to be stars but their low energy light emission (infrared) indicates that they are energy depleted, and so can no longer produce capacitor rings. Thus Brown Dwarfs are ageing stars or energy depleted stars or Stellar Cores.

In conclusion, Brown Dwarfs have strong magnetic fields like stars because they are old or energy depleted stars, or Stellar Cores, and not sub-stellar objects. This agrees with the old theory of stellar evolution in which Brown Dwarfs were ageing stars that had lost the ability to emit visible light, could only emit infrared light and were on their way to becoming black dwarfs, at which stage they would not emit any light at all. This view of Brown Dwarfs has been scrubbed from the internet. The reason for this appears to be that 'the powers that be' are terrified that the scientific community and the earth's population will find out that we have a system of Brown Dwarfs or energy depleted stars, in the inner Solar System, and that these objects are destroying the Sun and causing a slowly progressing or long-term cataclysmic event, on the Earth.

References:

[1]   Albers, C. (2018). Article 315: Rogue planet with a magnetic field 200 times stronger than Jupiter's.
[2]   https://www.newscientist.com/article/2147636-brown-dwarfs-have-strong-magnetic-fields-just-like-real-stars/
[3]   Albers, C. (2018). Article 146: Planet X System: Time of arrival.
[4]   Albers, C. (2018). Article 244: The Planet X System: destroyer of Star Systems (in Book 7: Planet X The effects on the Earth and the Sun).
[5]   Albers, C. (2018). Article 132: Survey of the Orion Nebula: Jupiter and Saturn may be stars.
[6]   Albers, C. (2018). Article 169: Planetary formation: comets to planets (in Book 3: Planet X revealed Gravity and Light).

Chapter 9

323. Planet X System Gravity: why has the earth not been destroyed?

Dr. Claudia Albers, Planet X Physiciat

I have been writing for a very long time that the Planet X System Objects, which I also refer to as Stellar Cores, do not obey the expected gravitational laws. This is because the strength of the gravitational interaction, embodied in the gravitational constant *G* is dependent on the photon energy contained in the particles making up an object (see Article 182: Einstein's dream realized: unified field theory of electrogravitation and Article 210: Stellar Core gravity: tidal and *G* is not constant) [1, 2]. The largest Stellar Cores are dead stars and thus depleted in energy and therefore can only interact through a very weak gravitational interaction. The Planet X System is actually a system of Black and Brown Dwarfs. Brown Dwarfs are in fact also old and thus energy depleted stars and on the way to becoming Black Dwarfs. Brown Dwarfs are not substellar objects (see Article 317: Planet X System: Brown and Black Dwarfs with the high magnetic field) [3]. Brown Dwarfs are not as energy depleted as Black Dwarfs as Brown Dwarfs emit infrared radiation and Black Dwarfs emit no radiation at all but the term Stellar Cores encompasses both.

Figure 9.1. Left: SDO composite image from July 31st, 2018 showing a Stellar Core make its energy drawing root-like connection to the Sun. **Right:** Striped Stellar Core moves through the corona drawing a trail of coronal plasma behind it. The stripes are characteristic of Brown Dwarfs which are in fact old energy depleted stars and the same object as Stellar Cores.

Now, to make the fact that these objects do not obey the standard way that gravity is understood to act, in this article, I am going to calculate the effect these objects would have had on Earth's orbit had they been able to interact gravitationally, with the same strength as Solar System objects. The Stellar Cores in the above images are small in comparison with some of the largest Stellar Cores that have been observed in the Sun's corona. The striped Stellar Core is one 4 times larger than the Earth, whilst the one in the left SDO image is about 8 times larger than the Earth and thus only a little smaller than Jupiter. But there are Stellar Cores that are several times larger than the Sun (see Article 321: Huge Planet X star in the inner Solar System) [4]. But in this article, we are going to start with one that is only about the same size as the Sun and determine what the effect it should have had on the earth's orbit had it been able to interact gravitationally like normal Solar System object.

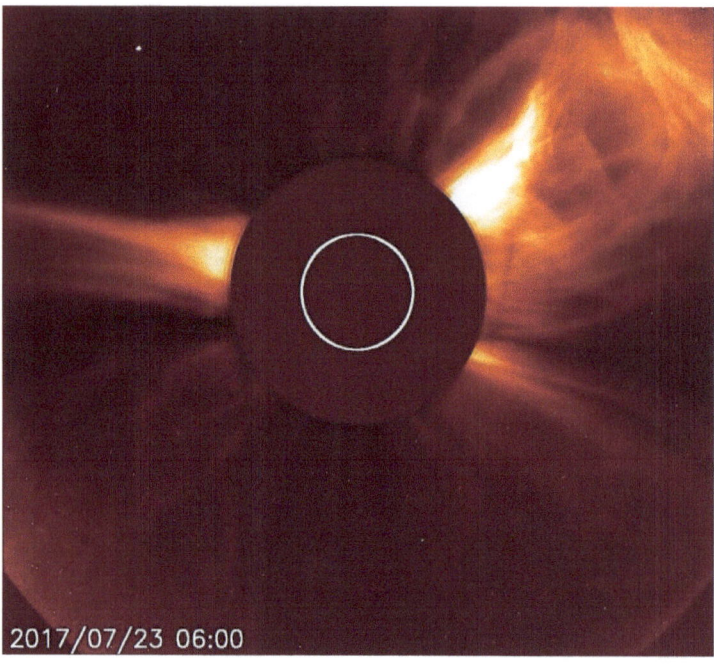

Figure 9.2. Stellar Core in a LASCO C2 image from July 23^{rd,} 2017 moving away from the Sun within a CME: It must be within the Sun's outer corona. A size comparison with the Sun reveals that it must be about the same size as the Sun.

Now, Stellar Cores shed their outer layers of material until their cores are exposed and will, therefore, have been larger than they are now as living stars. However, the core will undergo expansion to about twice its size, which will decrease its density but it should still be denser than the overall object would have been before it lost all its outer layers of material. This means that a Stellar Core, with the same radius as the Sun, should be denser than the Sun, and thus more massive, but we are going to assume that the Stellar Core in the above image has the same mass as the Sun. For this calculation, we are going to use Newton's Universal law of Gravitation and use conservation of energy. A planet's gravitational potential energy is given by

$$U = -\frac{GmM}{r} \tag{1}$$

where G is the gravitational constant, m is the mass of the planet, M is the mass of the Sun and r is the earth's distance from the sun. The kinetic energy of a planet, in an elliptical orbit, at a distance r from the Sun is given by

$$K = \frac{1}{2}mv^2 = \frac{GmM}{2r} \tag{2}$$

Then the mechanical energy, which is the sum of the kinetic and potential energies can be written in terms of the semi-major axis of the orbit

$$E = K + U = -\frac{GmM}{2a} \tag{3}$$

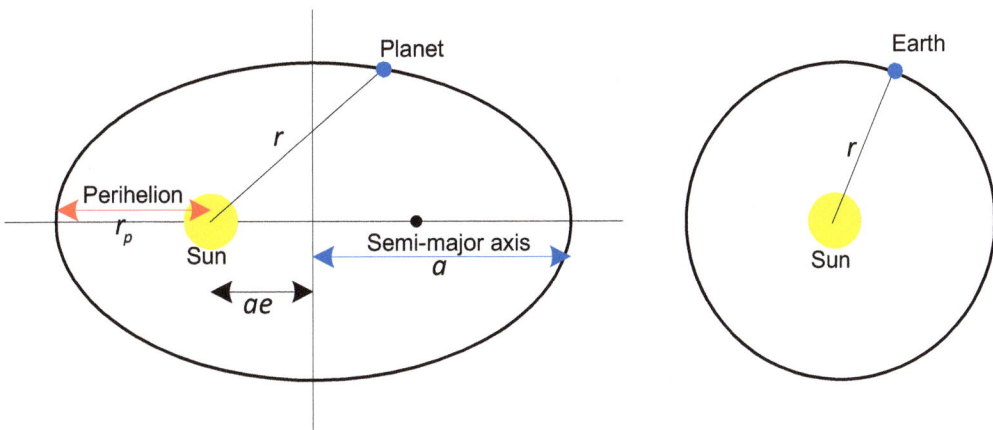

Figure 9.3. The semi-major axis, a, is the distance from the center of the ellipse to the perimeter, along with the longest diameter, e is the eccentricity. When e = 0, the ellipse becomes a circle of radius a. The earth's orbit is close to circular so $r = a = 1$ au (astronomical units).

Substituting equations (1) and (2) into equation (3) we obtain:

$$K + U = \frac{1}{2}mv^2 - \frac{GmM}{r} = -\frac{GmM}{2a} \tag{4}$$

Now, if we have another star arriving at the Sun, with the same mass as the Sun, this is the same as having a sun with twice the mass of the Sun, which will change the equation to:

$$\frac{1}{2}mv^2 - \frac{Gm2M}{r} = -\frac{Gm2M}{2a} \tag{5}$$

Then substituting for the kinetic energy from equation (1), we get:

$$\frac{GmM}{2r} - \frac{Gm2M}{r} = -\frac{Gm2M}{2a} \Rightarrow -\frac{3}{2}\frac{1}{r} = -\frac{1}{a} \Rightarrow a = \frac{2}{3}r = 0.67 \text{ au} \tag{6}$$

This means that the earth will now have a highly elliptical orbit with a semi-major axis of 0.67 au. Now, Venus has an orbital radius of 0.72 au, so this new semi-major axis is less than that. This means that the earth would then be closer to the Sun than Venus.

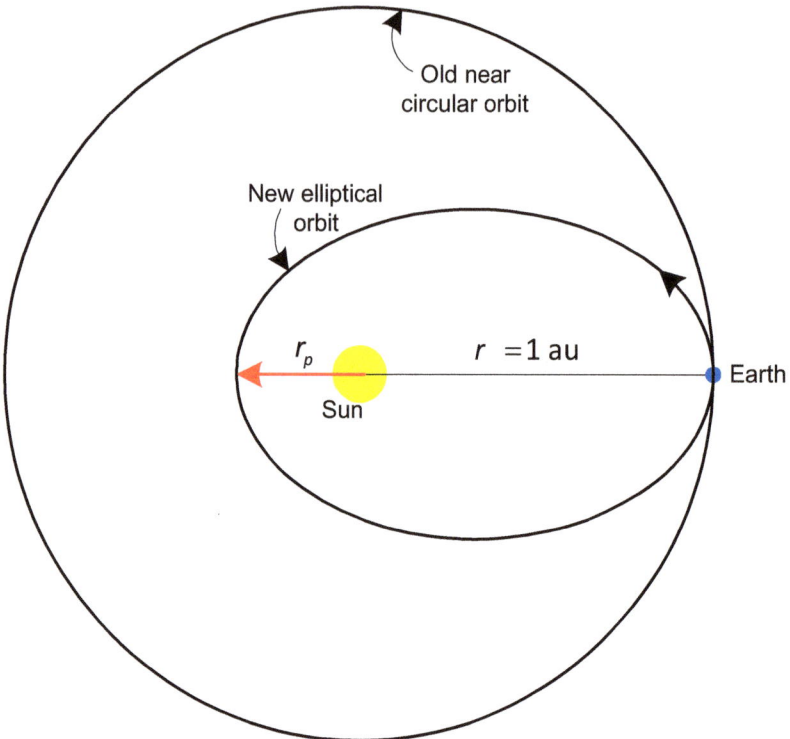

Figure 9.4. With another star with the same mass as the Sun very close to it, the effect should be the same as doubling the Sun's mass. This causes the Earth's orbit to become elliptical with a semi-major axis of 0.67 au.

Now, the total distance from one side of the ellipse to the other along the diameter is $2a$, so we can calculate the Earth's new perihelion position, or the closest distance it will get to the Sun, as it moves in its new orbit to be:

$$r_p + r = 2a \implies r_p = 2a - r = 2\left(\tfrac{2}{3}r\right) - r = \tfrac{1}{3}r = 0.33 \text{ au} \tag{7}$$

This perihelion position is less than Mercury's average orbital distance to the Sun, which is 0.39 au, and just a little over Mercury's perihelion position, which is 0.31 au. This type of disruption to earth's orbit would be catastrophic. But we have objects that are even larger, such as the Stellar, with 4 times the radius of the Sun. So next, we are going to assume that it has 4 times the mass of the Sun. In this way, its effect, if it was able to interact gravitationally in the same way as known Solar System object, would be the same as the Sun's mass suddenly increasing by a factor of 5.

Figure 9.5. Huge Stellar Core within a CME in a Stereo COR2 image from September 13[th,] 2017 at 7:11 (UTC). The object was initially estimated to be at least 3 times larger than the Sun, but a more careful analysis revealed that it was at least 4 times larger than the Sun (see Article 265: Planet X Object 3 times the size of the Sun) [5].

So, in this case, the earth's new orbit would have a new semi-major axis, a, given by:

$$\frac{GmM}{2r} - \frac{Gm5M}{r} = -\frac{Gm5M}{2a} \Rightarrow -\frac{9}{2}\frac{1}{r} = -\frac{5}{2a} \Rightarrow a = \frac{5}{9}r = 0.56 \text{ au} \quad (8)$$

And the earth's new perihelion position would be:

$$r_p + r = 2a \Rightarrow r_p = 2a - r = 2(0.56 \text{ au}) - 1 \text{ au} = 0.11 \text{ au} \quad (9)$$

Thus, the earth would come extremely close to the Sun and therefore its new orbit would be even more catastrophic. With an object, with 7 times the mass of the Sun close to it, and we have Stellar Cores of that size, then Earth's new trajectory would take into a collision course with the Sun, and it would not survive at all.

In conclusion, the Stellar Cores that are observed very close to the Sun cannot have a normal gravitational influence, because if they did we would not be here, anymore. The only possible reason why the Earth and the other planets in the Solar System have not been destroyed is that the Stellar Cores, which are really Brown and Black Dwarfs, have very low gravitational influence, so that the gravitational constant *G*, for the interaction between them and any other object, is close to zero. These objects have low gravitational influence because they are old and spent stars, and thus low in gravitational energy. This makes them into energy black holes; they suck photon energy from any object around them that has more energy than they do. It is this energy drawing mechanism that makes this system of Brown and Black Dwarfs, which have come from outside the Solar System, into an existential threat to Earth.

References:

[1]   Albers, C. (2018). Article 182: Einstein's dream realized: a unified field theory of electrogravitation (in Book 3: Planet X revealed Gravity and Light).

[2]   Albers, C. (2018). Article 210: Stellar Core gravity: tidal and G is not constant (in Book 6: Planet X Physicist Articles Part 1).

[3]   Albers, C. (2018). Article 317: Planet X System: Brown and Black Dwarf with the high magnetic field.

[4]   Albers, C. (2018). Article 321: Huge Planet X star in the inner Solar System.

[5]   Albers, C. (2018). Article 265: Planet X Object 3 times the size of the Sun.

Chapter 10

324. Planet X causes the sun to be darkened

When Jesus said in Matthew 24:29: '*²⁹ Immediately after the tribulation of those days shall the sun be darkened, ..*' he meant just that because the Sun is now being darkened, in other words, the Sun is literally going dark. In this article, I will detail how I discovered that the Sun does go dark.

When I started investigating the Planet X phenomenon, I did so because I saw pink colored clouds, in the sky, and I knew that it was impossible that such a color could be observed in the sky unless there was a source of the light other than the Sun emitting this color of light. What I discovered was beyond anything that I could imagine would even be possible. Observations in the form of images showing the objects in the Sun's corona do not lie. These objects are real and the implications are astounding as they change just about everything I was taught. What I learnt has led me to discover what causes gravity, a question that I was often asked by my students, when I was still lecturing at the University of the Witwatersrand, in South Africa. Gravity comes from the photon or light as does everything else in the universe and the strength of the gravitational interaction is determined by light, or photon, energy, inside the particles of an object (see Article 190: The God particle: Turning light into matter and Article 210: Stellar Core gravity: tidal and *G* is not constant) [1, 2].

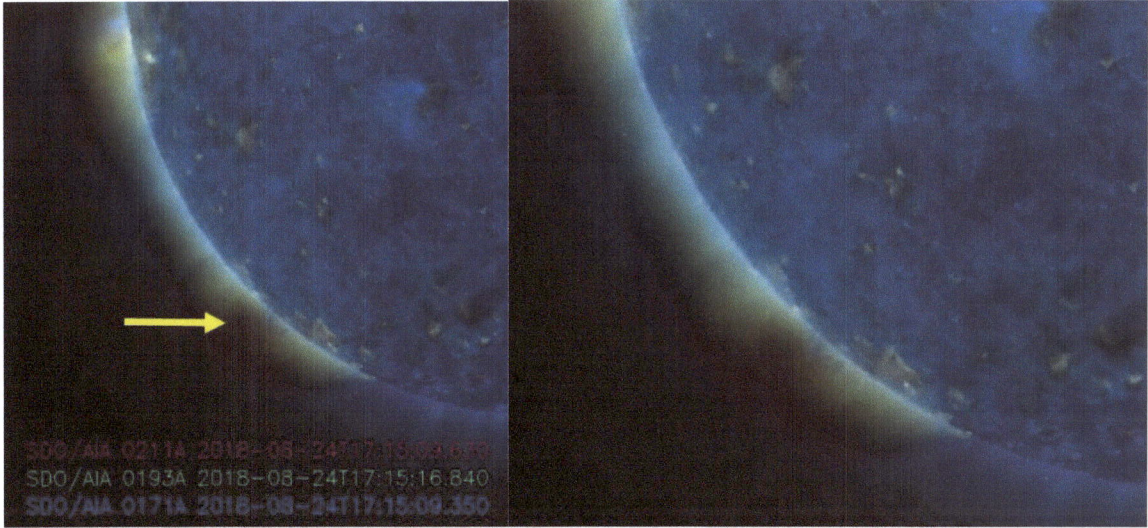

Figure 10.1. Composite SDO images from August 24th, 2018 showing a Stellar Core in the Sun's corona (caught by Scott C'one).

But on the way to discovering the truth about gravity, I first discovered that the Sun was going dark at certain times. The fact that it was going dark could be seen in the images obtained by the SDO satellite during the so called SDO eclipse season, which is supposed to occur twice per year. From the first time I saw these images showing the Sun going dark, I knew that only the Sun actually going dark could actually produce these types of images. These could not have been produced by an eclipse and therefore the eclipse has to be a cover story. The realization that the Sun was capable of going dark was a huge step in my understanding of what was going on in the Solar System, and of how the universe really works. For one thing, it means that just about all currently accepted astrophysics theory, which is based on stars being powered by thermonuclear reactions and on gravitational collapse, is completely wrong. This is because there cannot be any thermonuclear reactions occurring in the Sun, as if there were, there would be light continuously coming from inside the Sun, and it would be impossible for the Sun to go dark. Thus, the understanding that the Sun is going dark is of extreme importance in understanding how the universe really works. I have written about this many times, but some people still find the arguments difficult to understand, so in this article, I am going to attempt to explain it more clearly.

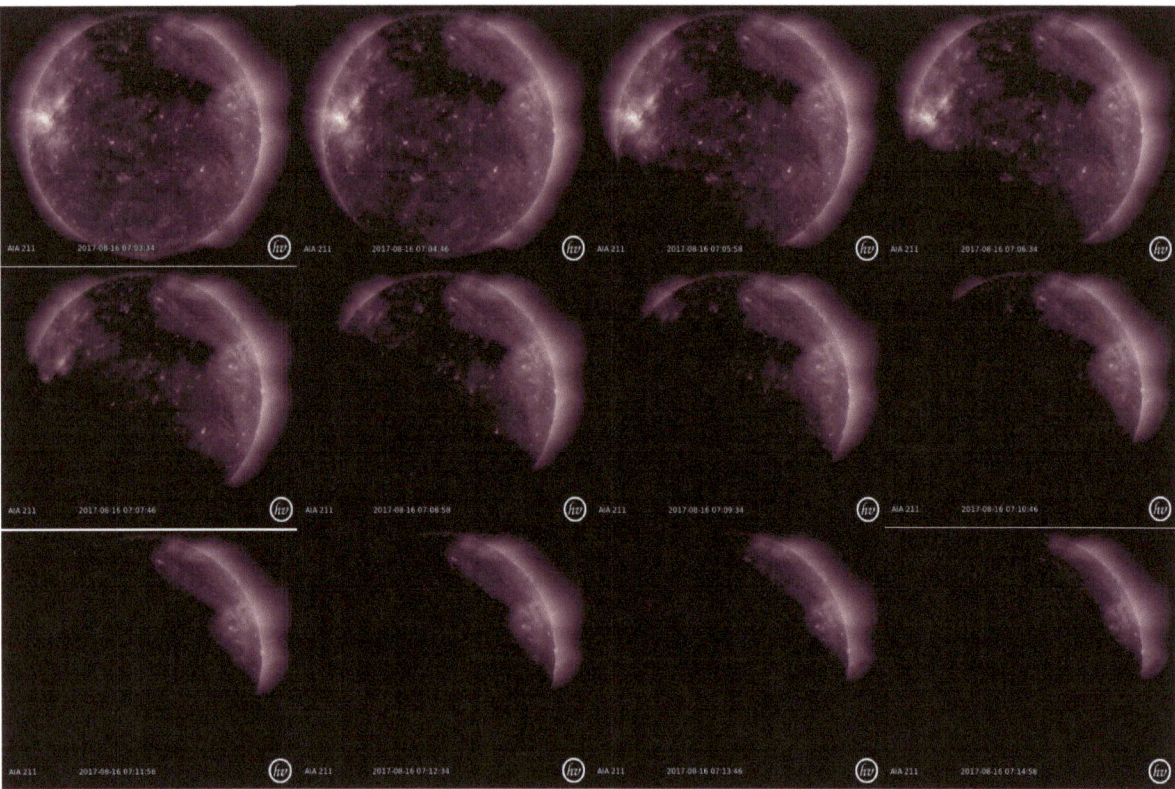

Figure 10.2. Images of the Sun as provided by the SDO satellite, from August 16th, 2017, in the 21.1 nm wavelength, between 7:03 and 7:14 (UTC). Completed solar structures are seen at the edge of the dark/light interface. The earth does not seem to be partially covering anything. This and the shrinking corona shows that the Sun is not being eclipsed but is actually going dark.

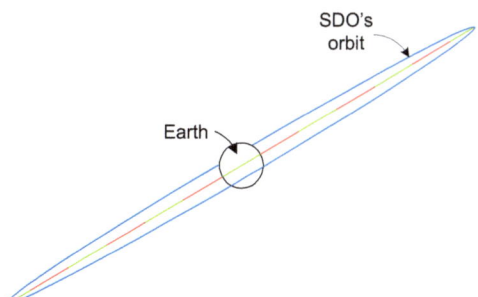

Figure 10.3. SDO's orbit and the Earth to scale: The period is 24 hours (geosynchronous with a 28° orbital inclination) so the Earth can only block the satellite's view of the Sun twice per year.

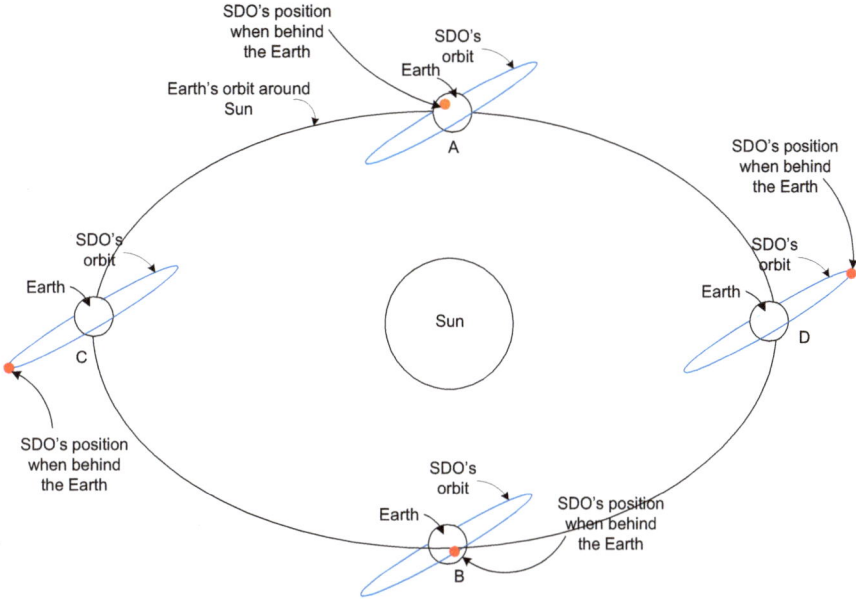

Figure 10.4. SDO's position relative to the Earth's position at 4 different times during Earth's orbit around the Sun. At positions C and D, SDO gets an uninterrupted view of the Sun, as the Earth would be above (at C) or below (at D) the Sun, from SDO's view point, when it goes behind the Earth. At positions A and B, the Earth eclipses the Sun, from SDO's perspective.

Then, using the fact that Earth's radius is 6 371 km, the sun's radius is 696 300 km, and the distance, between the Earth and the Sun, is 152 million km, we can calculate the Sun's angular width to be:

$$\theta_{Sun} = \frac{2R}{d} = \frac{2(6.963 \times 10^5)}{1.52 \times 10^8} \text{ rad} \left(\frac{180°}{\pi}\right) = 0.52°$$

and the Earth's angular width, as viewed from SDO's position, is then:

$$\theta_{Earth} = \frac{2r}{r+h} = \frac{2(6.371 \times 10^3)}{4.21 \times 10^4} \text{ rad} \left(\frac{180°}{\pi}\right) = 17.3°$$

Thus, from SDO's point of view, the Earth will seem to be 33 times larger than the Sun.

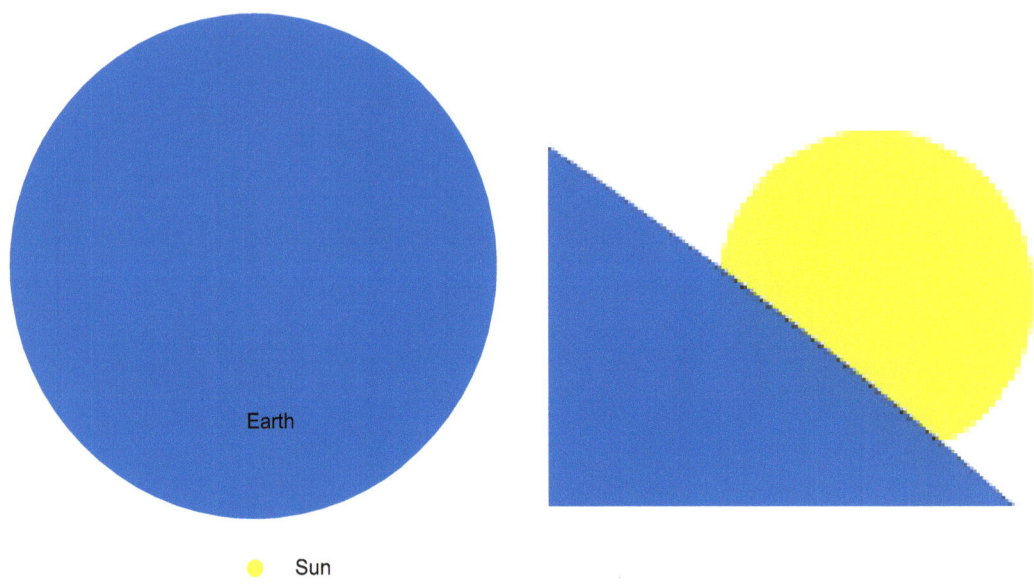

Figure 10.5. Left: The Earth's size in comparison with the Sun's, from SDO's perspective. **Right:** The earth's curvature in comparison with the Sun's is only slightly positive or almost flat (infinite) so we can use a triangle to represent the Earth.

One of the arguments that a colleague of mine, who is an astronomer, gave me in order to discourage my stand on the SDO eclipse season, was that part of the Sun can be viewed through the earth's atmosphere. However, the earth's atmosphere absorbs all extreme ultraviolet and x–ray wavelengths, coming from the Sun. SDO detects several extreme ultraviolet, one x-ray, and one visible light wavelength, so only one wavelength, the visible light one, can be affected by the satellite viewing the sun through the atmosphere. This is why a satellite is used to view the Sun in extreme UV, these cannot be viewed from within the atmosphere. In addition, even though up to a third of the Sun could technically be viewed through the earth's atmosphere, it is obvious that features which are bright, within that one third, remain equally bright in subsequent images until that section goes dark.

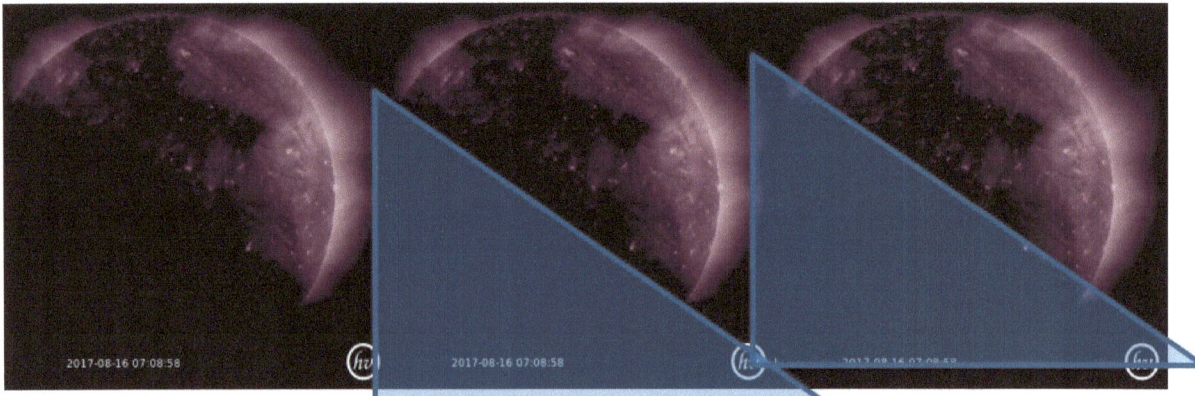

Figure 610. Left: SDO image where about half of the Sun is dark. **Center and right:** The earth is represented by the blue triangle. Where would the earth be? Below where the corona ends and becomes flush with the face of the Sun is the more likely position, because if further up, the lower part of the corona would be covered by the earth, and thus not visible.

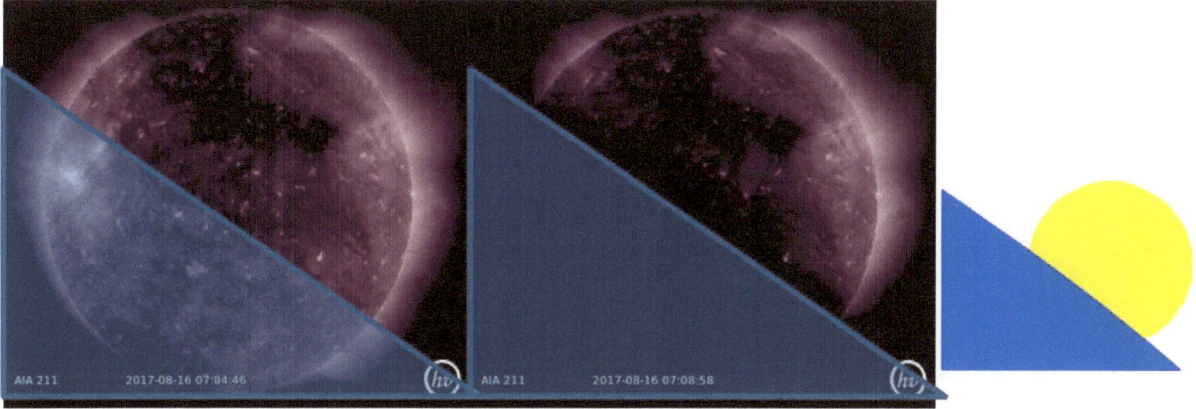

Figure 10.7. When we compare the image on left, in figure 6, with one from 4 minutes before, we see that the corona is not covered, but it has shrunk. The earth is not curved enough to produce the darkness seen above the triangle. And, how can several cellulars like parts of the Sun remain visible below the edge of the darkness, if the earth is supposed to be covering them?

The Sun does not seem to have been covered by the Earth at all; it is as if a coronal hole is spreading across the face of the Sun. Notice also how the top coronal hole seems to have grown and spread below the original coronal hole. This clearly indicates that the Sun is going dark, and it is not going dark due to an eclipse but rather due to loss of light emission.

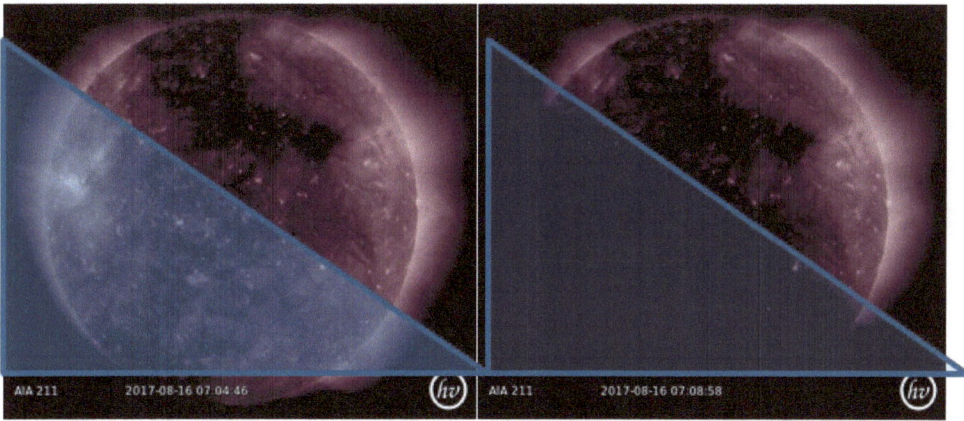

Figure 10.8. Even with the earth positioned further up, it is obvious that the corona is not covered but has shrunk indicating that the Sun is going dark. The earth cannot be in this position though, as then several features seen in this image would be covered by the earth, again showing that there is no earth between the satellite and the Sun and this cannot, therefore, be due to an eclipse. This is the sun progressively losing light emission.

This loss of light emission occurs at all wavelengths detected by the SDO satellite as shown in figure 9 below.

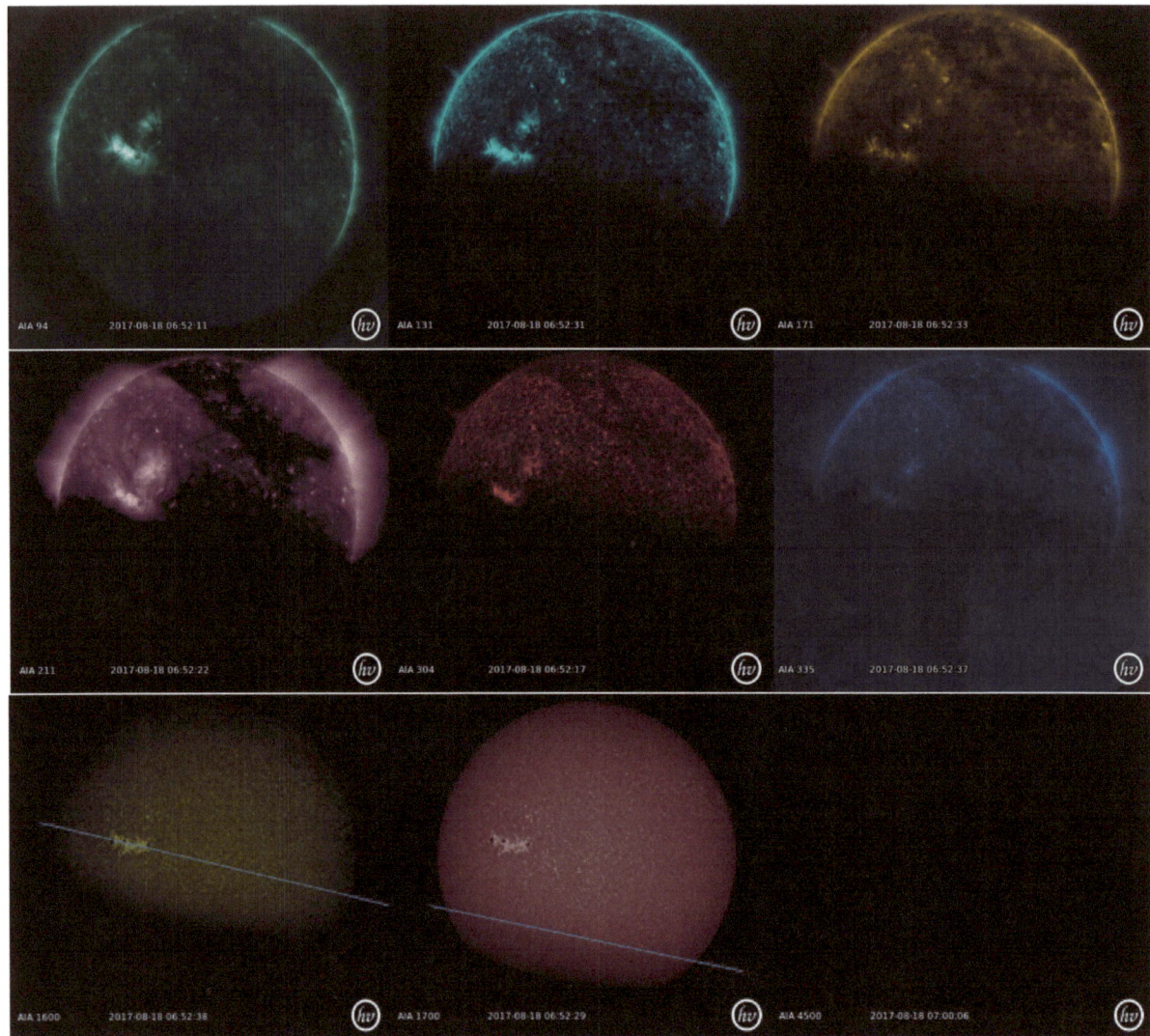

Figure 10.9. Images of the Sun coming from the SDO satellite, from August 18[th,] 2017 at 6:52 (UTC) in several wavelengths: The last image is in the 450 nm (visible light) wavelength and is from 7:00 (UTC). The Sun had gone completely dark by then. This shows that the Sun's light emission is affected at all wavelengths, including visible light. Different portions of the Sun are visible at the different wavelengths showing that the Sun loses light emission at different rates, for different wavelengths. It also seems to lose visible light emission, before losing light emission at the other wavelengths. The earth would have to have a negative curvature to produce the 1600 and 1700 angstrom images, which is impossible.

It is therefore impossible for the SDO images, produced during the SDO eclipse season, to be due to an eclipse. These images can only be produced if the Sun is actually going dark at this time. But how would the SDO satellite not actually observe a real eclipse twice a year, as a result of its orbit causing it to be behind the earth? I think, it is likely that there are, at least, two SDO satellites up there, and when the Sun is being eclipsed for one, the other is still able to provide images of the Sun.

So what causes the Sun to lose light emission? The Sun produces light as a result of a high electric potential across it atmosphere but since the Stellar Cores (Brown and Black Dwarfs [3]) belonging to the

Planet X System absorb electrons and drain the Sun of energy, they are able to cause a drop in the Sun's electric potential when they closely approach the Sun. This also shows that the Sun's surface cannot be any hotter than 750 K (900°F) (see Article 195: Stellar Cores and the dying Sun and Article 232: The Sun can go dark: the implications) [4, 5].

Now, the SDO satellite was officially launched on February 11th, 2010 and the SDO eclipse season occurs in February/March and in August/September, but in 2010 there was no eclipse season in March. So the first eclipse season was in September of 2010. Images from this eclipse season, in 335 angstroms, are shown below.

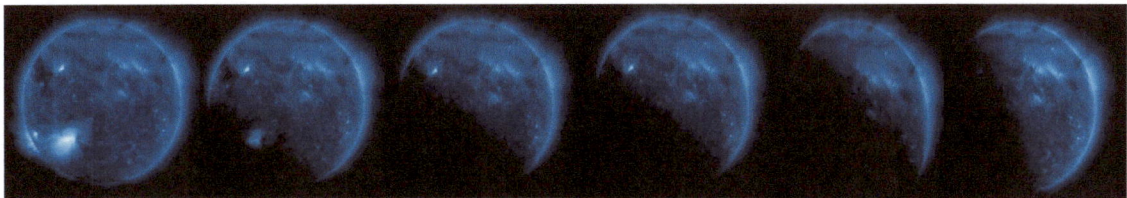

Figure 10.10. Images of the Sun as detected by SDO in 335 angstroms (ultraviolet) between 6:40 and 7:07 (UTC) on September 15th, 2010.

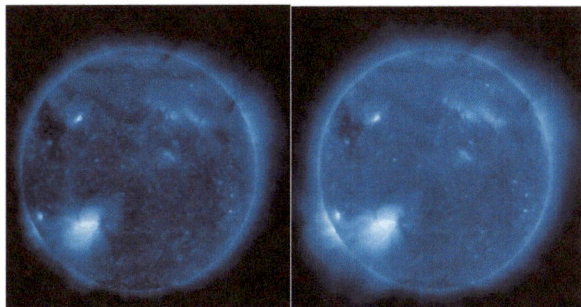

Figure 10.11. Images of the Sun as detected by the SDO satellite in 335 angstroms (ultraviolet) at 7:07 and 7:08 (UTC) on September 15, 2010, just as the Sun regains light emission after having gone dark. There is a marked increase in the Sun's brightness between the two images. Also, the whole corona grows in size, indicating that a shrinking corona is associated with loss of light emission. Notice how the small coronal hole, on the left side of the Sun, also decreases in size.

In conclusion, the Sun can, and does, go dark. The fact that the Sun's corona shrinks back as darkness progresses across the face of the Sun provides irrefutable evidence that the SDO eclipse season images are not due to an eclipse but actually show the Sun going dark. This irrefutable observation makes all the current accepted theories about how the Sun operates and how stars form since they are all based on gravitational collapse, completely wrong. It took the invasion of the Planet X System of our Solar System to show us that everything we have been led to believe about the universe is wrong.

References:

[1] Albers, C. (2018). Article 190: The God particle: Turning light into matter (in Book 6: Planet X Physicist Articles Part 1).

[2]   Albers, C. (2018). Article 210: Stellar Core gravity: tidal and *G* is not constant (in Book 6: Planet X Physicist Articles Part 1).
[4]   Albers, C. (2018). Article 195: Stellar Cores and the dying Sun (in Book 6: Planet X Physicist Articles Part 1).
[4]   Albers, C. (2018). Article 317: Planet X System: Brown and Black Dwarf with high magnetic field.
[5]   Albers, C. (2018). Article 232: The Sun can go dark: the implications (in Book 6: Planet X Physicist Articles Part 1).

Chapter 11

336. Stellar nebular cloud structure

In Article 169: Planetary formation: comets to planets [1], I wrote about the solar capacitor: a series of rings or nebular clouds centered on a star, which is produced from the star's production of solar wind. The concept came from James McCanney, who in his book; Planet X Comets and Earth Changes [2] describes the structure of these nebular clouds in terms of capacitor plates and calls them the solar capacitor. According to him, these nebular clouds are made out of positive ions, which the Sun ejects as solar wind, with the heaviest remaining in the cloud closest to the Sun, and the lightest moving to the furthermost cloud from the Sun. He also states that that the negative plate is the Sun's mainly electron outer layer or corona. He deduced this structure from observing the behavior of comets and how they seemed to absorb the material, found in their tail, from their environment.

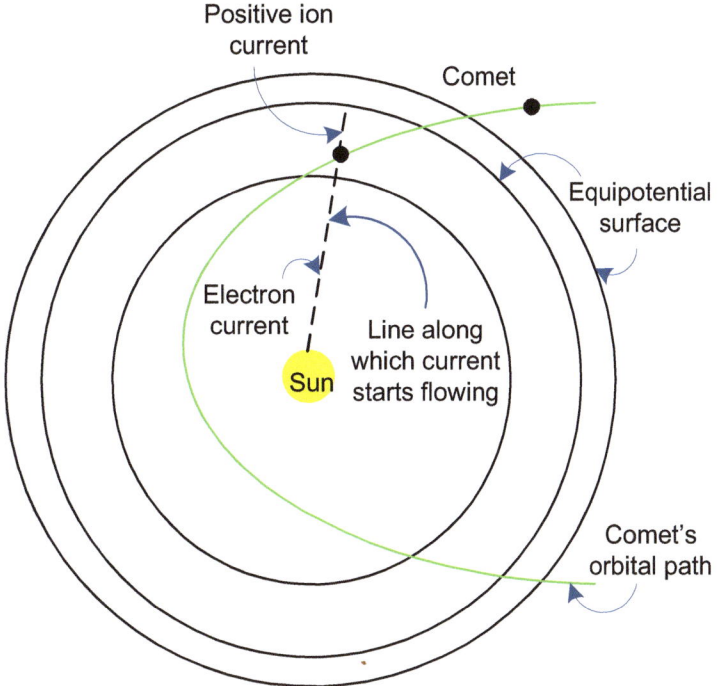

Figure 11.1. The Solar Capacitor structure of nebular clouds, around the Sun according to James McCanney: The nebular clouds are made of positive ions (atoms with missing electrons) and thus form positive capacitor plates. A comet coming into the Solar System absorbs ions from the closest nebular cloud and electrons from the Sun's negative plate or outer negative layer or corona.

All stars, which still retain enough energy generation power, in their cores, to produce a solar wind, will have either a nebular cloud structure or a ring structure, around them, and centered on the star. Since Jupiter has nebular clouds, and Saturn has a ring structure, they should, therefore, both be viewed as stars, not planets.

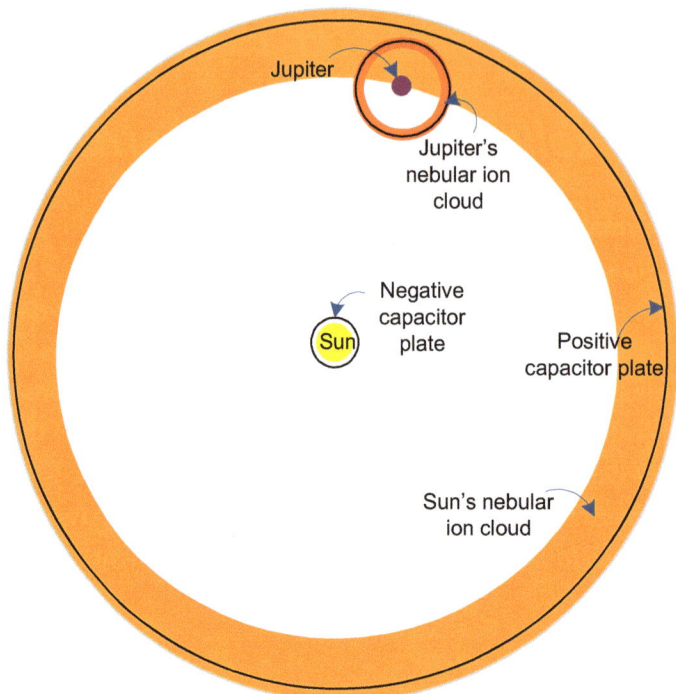

Figure 11.2. Jupiter's nebular cloud within the Solar Capacitor: When comet Shoemaker-Levy 9 passed into the region of Jupiter's capacitor it started drawing in, through its tail, sulfur, and oxygen, instead of hydrogen and oxygen, and thus the tail went from containing water (H_2O) to containing sulfur dioxide (SO_2). This showed that Jupiter's nebular clouds contain sulfur and oxygen, rather than hydrogen and oxygen, contained in the Sun's nebular clouds, at Jupiter's distance from the Sun.

Young and very energetic stars are likely to have the largest nebular cloud structure, which may also emit light, which is the reason why so many stars have what is called planetary nebulae. These planetary nebulae are usually attributed to being the result of a star has gone through a supernova, at the end of its lifetime, so that the star at the center of the nebulae is viewed as a dead star. Actually, what we have is the exact opposite, only a young energetic star is able to have a strong stellar wind, which will then give rise to a large planetary nebula. Old stars, as we know, from observing the Stellar Cores, in the Sun's corona, are not able to emit light, or have a solar wind, because they are no longer able to generate energy, in their core (see Article 184: Stellar Core evolution) [3]. Thus, a supernova cannot signify the end of a star's life and is, therefore, most likely what occurs when a new star is born, possibly by fissioning, from another larger star.

Figure 11.3. Left: The Ring Nebula also referred to as M57. **Right:** The Helix Planetary nebula or NGC 7293. Planetary nebulae are characterized by a central star surrounded by a cloud of gases, which usually form a ring or toroidal pattern. The gases must be ionized and are thus the star's nebular clouds produced by the star's stellar wind, and are thus created as a result of the energy the star is able to generate, in its core, through radioactive decay. Only a young star will have the energy to produce such a beautiful nebular cloud structure (see Article 254: Planetary nebulae and death of a star) [4].

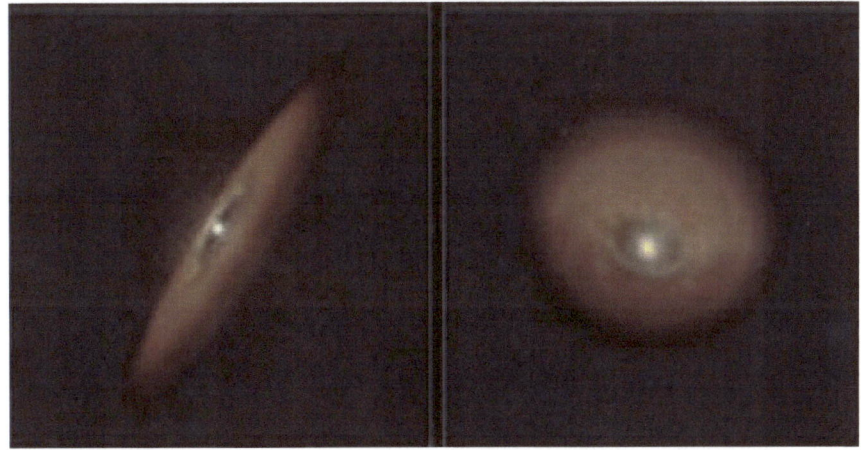

Figure 11.4. A zodiacal ring around a star: The ring closet to the star will contain the heaviest ions the star ejects such as iron and nickel

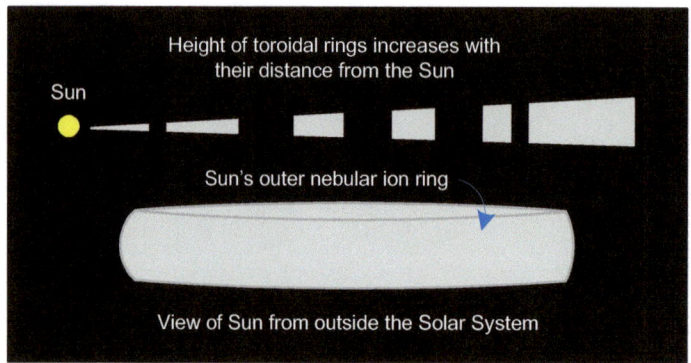

Figure 11.5. **Top**: illustration of how the toroidal rings or nebular clouds increase in height with distance from the Sun and are thus similar to the Earth's van Allen Belts. **Bottom**: View from outside the Solar System (from the ecliptic plane).

Although the structure proposed by James McCanney works very well to explain the phenomena surrounding comets, such as the production of a comet's tail and coma, the fact that the Earth's van Allen Belts, which are equivalent to the Sun's nebular clouds, are made of alternating positive and negative rings, with the negative ring on the outside, has made me realize that this is also the structure of the Sun's nebular clouds, i.e. there must be negative electron clouds interspersed by the ion clouds and that the last nebular cloud at the boundary of the Solar System and interstellar space must be a negative electron cloud.

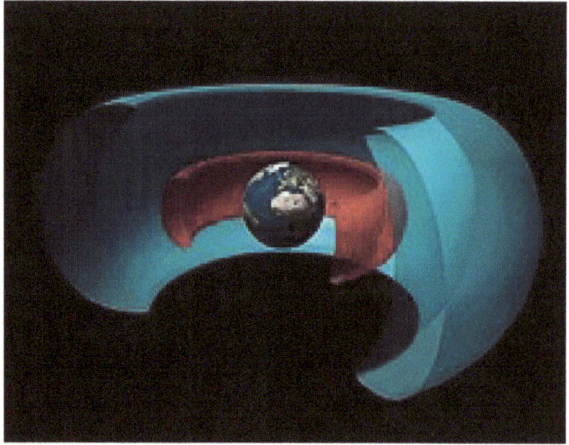

Figure 11.6. The earth's van Allen Belts: the inner one in red is filled with protons and the outer blue one is filled with electrons. Electrons always occur on the outside of isolated objects for the same reason that they are also found on the outer layers of an atom. This is a result of the gravitational interaction, which causes protons to strongly attract each other, and protons and electrons repel each other, but not as strongly as protons attract each other (see Book: Planet X Revealed Gravity and Light) [5].

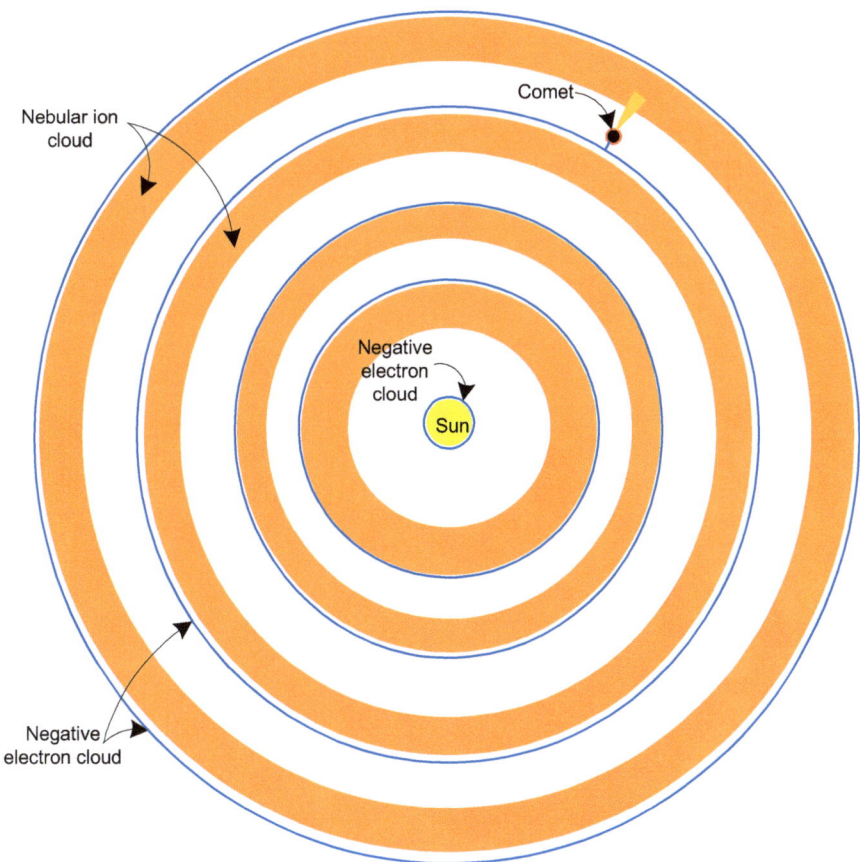

Figure 11.7. Solar Capacitor with nebular ion clouds and electron clouds: The comet absorbs ions from the nebular ion cloud behind it, and electrons from the electron cloud in front of it. The Solar System will be bounded by a final electron cloud.

The electron clouds are analogous to the energy levels of an atom, and the most energetic electrons will be in the outermost layer. These electrons will have more gravitational energy, in them, than electrons in the inner layers. The electrons are kept in the appropriate energy level, which is at an associated distance from the Sun, by the balance between the gravitational force generated by the Sun, in its core, and the electrostatic interaction between the particles in the core and the electrons. Thus, the whole Heliosphere, which is believed to go out to 100 au from the Sun, is the Sun's atmosphere, and all the planets, move within it. The planets' outer layer of electrons will have the gravitational energy appropriate for the distance, at which the planets orbit, from the Sun.

The nebular ion clouds are produced by the Sun's gravitational field, but indirectly. The gravitational interaction causes protons and electrons to repel and the higher the energy in the particles the stronger will be the repulsion, which is why electrons with more gravitational energy, end up further from the Sun. But the nebular ion clouds are ejected from the Sun, as a result of the electrical field generated in the Sun's outer layers, or the inner corona of the Sun. The field results from electrons being repelled by the dense positively charged core. The electric potential rises steeply, in this region, i.e. the corona, and results in electric discharges, which lead to the emission of high energy photons, which then split into particles, fusion then occurs, and several heavier nuclei form, all the way up to iron. These may absorb electrons, but they have then ejected away from the Sun's corona, depending on where they form. These nuclei or ions (if they have acquired electrons) may be pulled into the Sun's interior, or ejected into space outside of the corona. If the matter creation event occurs closer to the top of the corona, they are repelled by this main electron layer, out into space, and become the Sun's Solar Wind, which then feeds the nebular ion clouds of the solar capacitor.

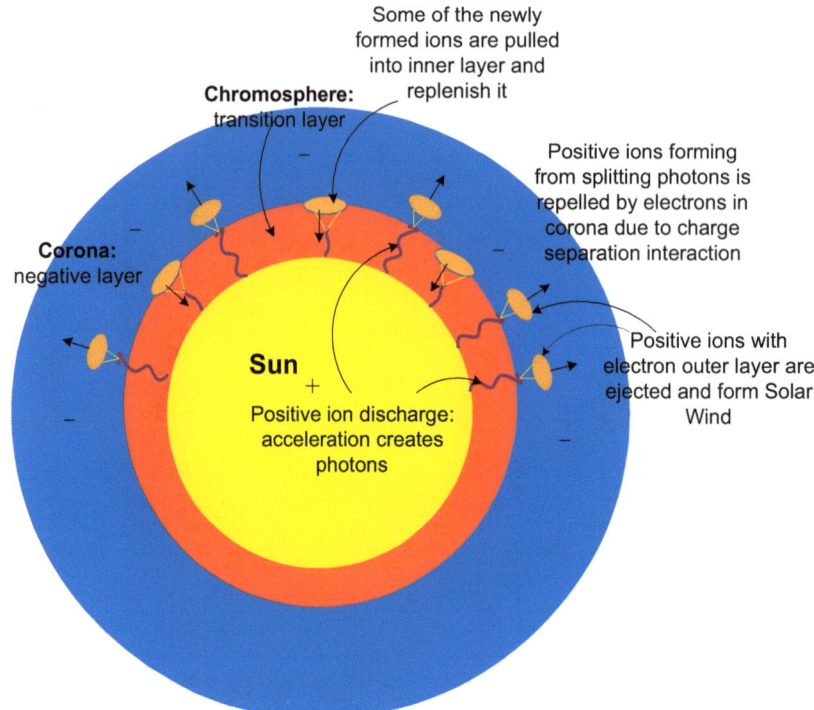

Figure 11.8. Illustration of how the Sun's Solar Wind, which feeds the Sun's nebular ion clouds, is produced. Before I realized that the gravitational interaction could be repulsive as well as attractive I called the repulsive part of the gravitational interaction, the charge separation interaction (see Article 175: How is the Solar Wind Produced? and Book: Planet X Revealed Gravity and Light) [6, 5].

Planets are not very different from stars. Planets also have electrical discharges in their atmosphere and thus also emit light from their atmosphere just like stars do. On earth, lightning discharges lead to the emission of gamma rays, which are energetic enough for positrons to appear [7]. Positrons are electrons with a positive charge and usually also referred to as anti-electrons. I, however, do not believe there any such thing as anti-matter, there are only particles, which emerge from within photons, with a certain mass, and which can have either a positive or a negative charge [5].

However, planets do not generate enough gravity through the energy generation process in their core to have a flux of ions from their atmosphere and thus produce their own nebular ion cloud. If a planet is able to generate its own nebular cloud or ring structure, then it should be classified as a star rather than a planet. However, planets also have nebular clouds or radiation belts with particles in them. The particles in the outer belts of planets seem to come from the Sun's Solar Wind, rather than from the planet's atmosphere. If there are any particles that come from the planet, such as high energy electrons these will join the solar wind electrons, at the appropriate energy level, or orbital distance, from the planet.

Nebular ion clouds have lower gravitational energy, the further we move from the star, at the center, as an ion's energy comes mainly from the number of protons in the nucleus and is thus mostly in the form of mass, but as we move further out from the star, lighter elements will populate the nebular clouds so the ions will have less mass and therefore less energy. But, the electron clouds, as we move outwards from the star, will be populated by electrons, with an increased amount of gravitational photon energy, i.e. photons inside the electrons which give it the ability to interact gravitationally [5].

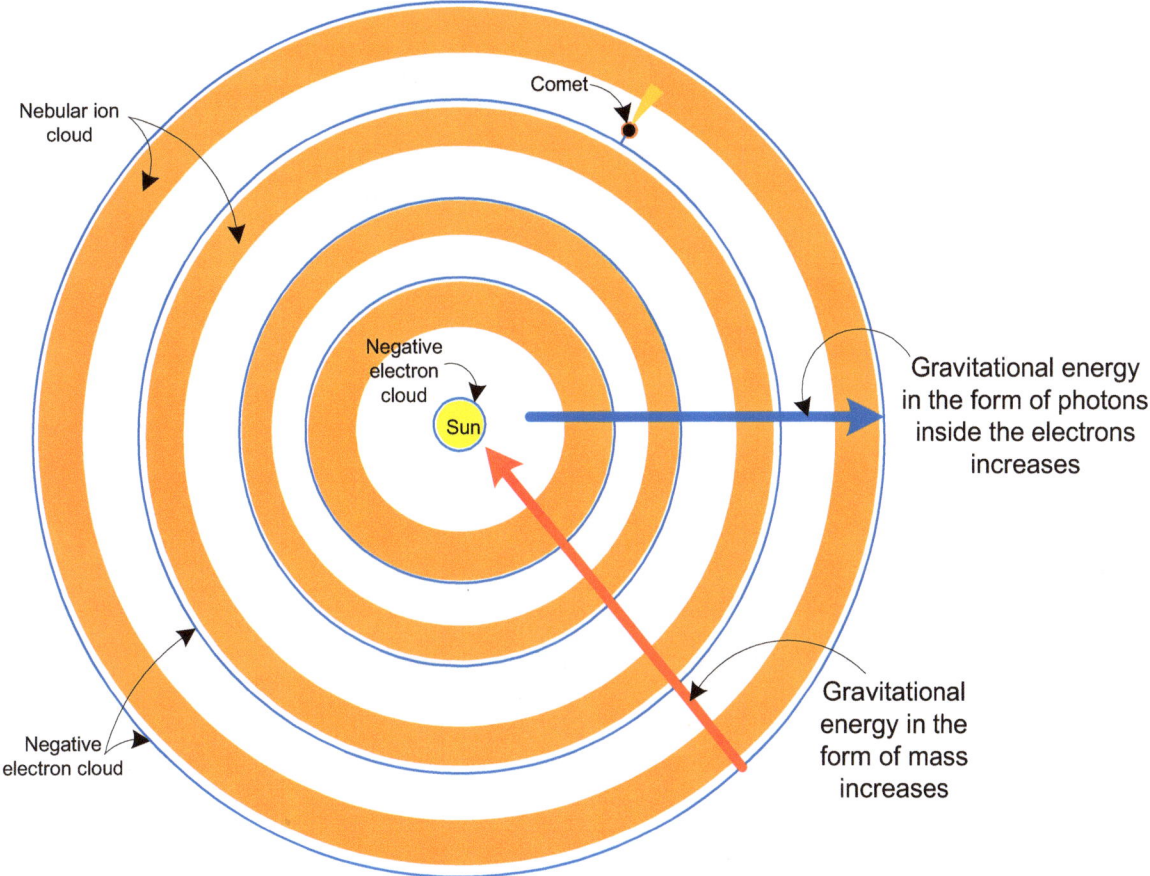

Figure 11.9. Electrons increase in energy as we move outwards from the Sun. This energy is in the form of gravitational energy or photons inside the electrons. Ions in the nebular clouds become heavier, i.e. have more protons in their nuclei as we inwards toward the sun so energy in the form of mass increases as we move inwards toward the Sun.

Mass and gravitational energy are two different forms of energy, which originate with photons or light, thus both are made out of light. When a photon splits into two particles of opposite charge, part of its energy turns into the mass of the particles, and the rest continues to exist as photon energy but inside the particles with mass. Thus, particles become containers for photons.

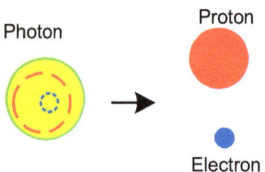

Figure 11.10. Photons split into constituent particles when moving through a region of high enough electric field. When this occurs, part of the photon's energy turns into mass, and the rest turns into gravitational photon energy, which exists within the particles.

In conclusion, the gravitational interaction leads to the Sun, and by extension, all stars, having a structure of nebular ion clouds and negative electron clouds. The gravitational energy of electrons in the electron clouds increases as we move outwards from the star just like the energy of the electrons, in an atom, and the ions, in the nebular ion clouds, increase in energy in the form of mass, as we move inwards toward the star.

References:

[1] Albers, C. (2018). Article 169: Planetary formation: comets to planets (in Book 3: Planet X revealed Gravity and Light).

[2] McCanney, J. (2002). Planet X Comets and Earth Changes. Jmccanneyscience.com press Minneapolis.

[3] Albers, C. (2018). Article 184: Stellar Core evolution (in Book 3: Planet X revealed Gravity and Light).

[4] Albers, C. (2018). Article 254: Planetary nebulae and death of a star.

[5] Albers, C. (2018). Article 175: How is the Solar Wind Produced? (in Book 3: Planet X revealed Gravity and Light)

[6] Albers, C. and C'one, S. (2018). Book 3: Planet X Revealed Gravity and Light. Amazon publishing.

[7] Enoto, T. et al. (2017). Photonuclear reactions triggered by a lightning discharge. Nature 551, pp. 481-484.

Chapter 12

337. Heat and gravitational photon energy

As detailed in Article 336: Stellar nebular cloud structure [1], photon energy can exist in three forms; as a free electron travelling at the speed of light, mass and as photon energy inside a particle. The last form is photon energy, which is contained within particles with mass. Furthermore, stars have nebular clouds of ions and electrons, which are fed by the stellar wind they produce. The electrons increase in energy the further they are from the star. In this way, these electron clouds are like the energy levels of an atom.

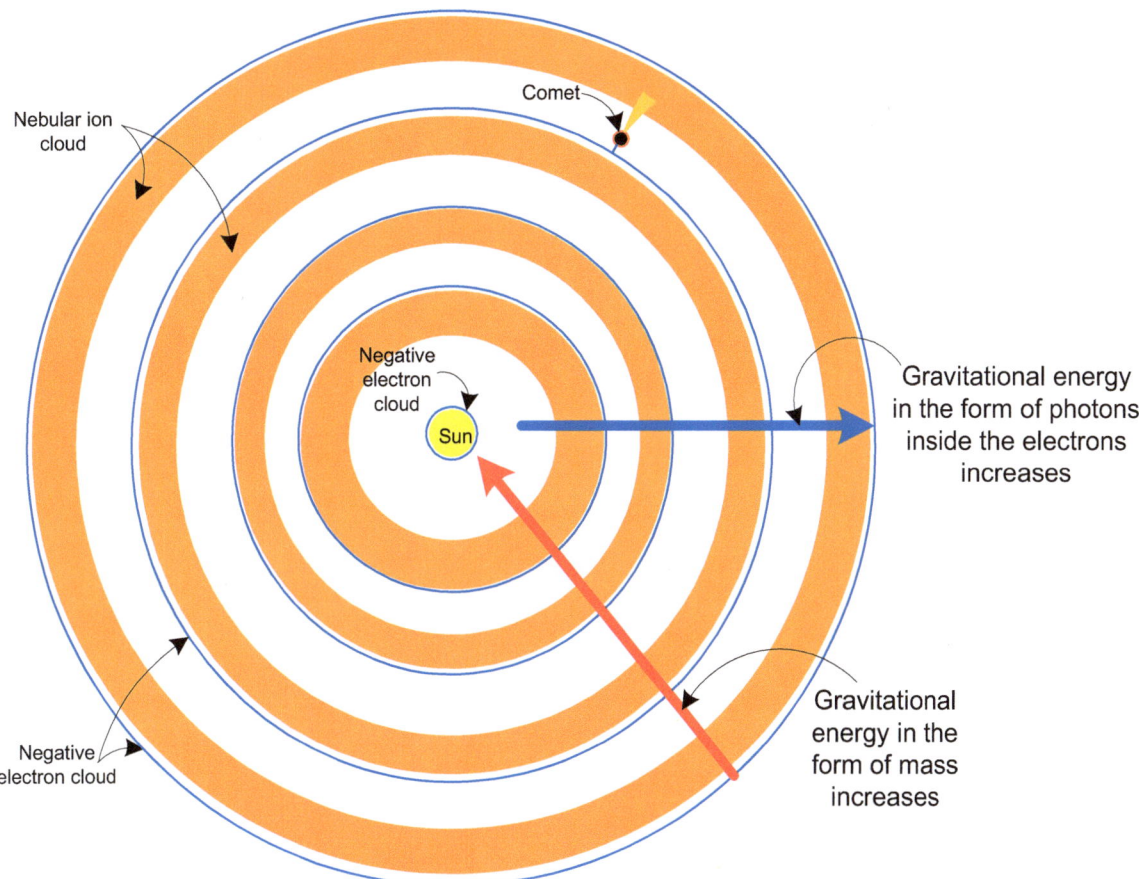

Figure 12.1. Electrons increase in energy as we move outwards from the Sun. This energy is in the form of gravitational energy or photons inside the electrons. Ions in the nebular clouds become heavier, i.e. have more protons, in their nuclei, as we move inwards toward the sun, so energy, in the form of mass, increases as we move inwards, toward the Sun. This is the same structure as in atoms where the most massive part is in the nucleus and electrons have higher energies as we move further outwards from the nucleus.

Planets also have radiation belts, called the van Allen Belts, in the case of Earth, which are actually analogous to the Sun's nebular clouds but seem to be populated by particles coming from the Solar Wind.

All celestial objects operate like atoms and are thus like superatoms.

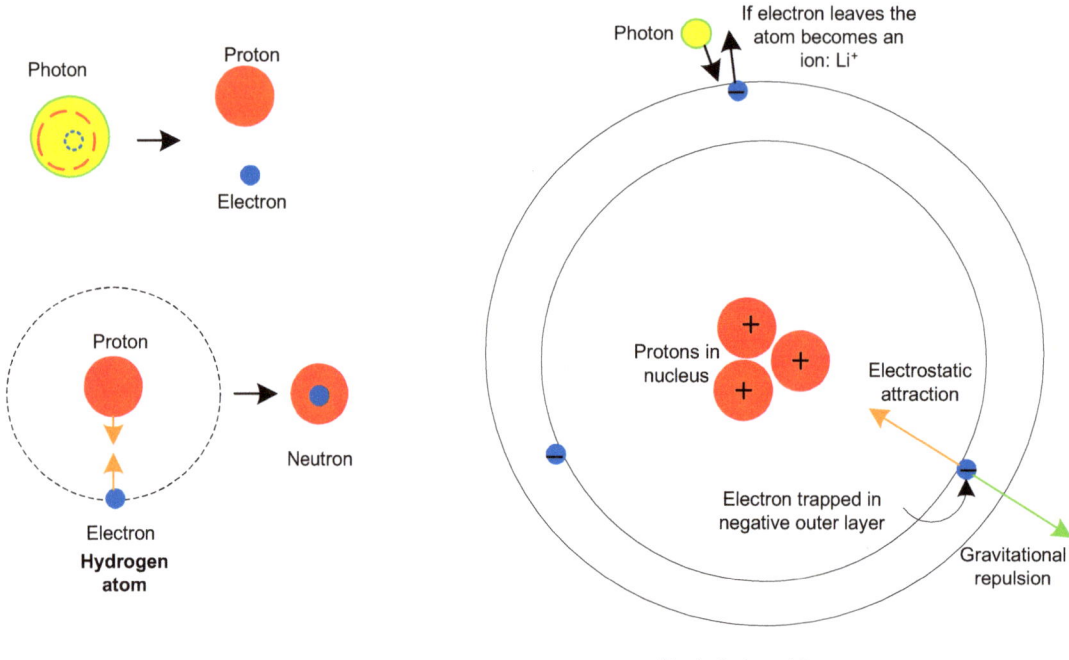

Figure 12.2. Photons split into their constituent particles. The gravitational interaction separates protons and electrons and attracts protons to protons so that neutral atoms can form. If the gravitational interaction is not strong enough these combine into neutrons. Electrons are trapped in an outer region of an atom, where the forces, exerted on the electron, due to the electrostatic and gravitational interactions, are in balance. Stars are atoms on the macroscopic scale.

Electrons can absorb photons, which carry gravitational energy, which thus increases the strength of the gravitational repulsion between the electrons and the nucleus, which may then result in the electrons leaving the atom, or moving further away from the nucleus. But, electrons can only have certain specific energies, within an atom, so, in order for an atom to move from one energy level to another, it needs to absorb a photon with a specific wavelength, which corresponds to the difference in the energy between the two energy levels. Thus, in order for a photon to be absorbed by an atom as the gravitational energy, it has to have specific energies or wavelengths.

However, any photon when incident on the matter will lead to an increase in its temperature. An increase in temperature is associated with an increase in heat energy, which is associated with an increase in kinetic energy. In other words, the absorption of photons leads to a particle moving faster or vibrating faster. Thus, when photon energy is incident on an atom, which does not correspond to the specific energy which will allow electrons to shift energy levels, this energy is absorbed as heat, and not gravitational energy. This, therefore, shows that photon energy, which can be contained within particles, is in two different forms: heat photon energy and gravitational photon energy.

The reason why there are only specific allowed energy levels, which correspond to only certain allowed orbits, seems to be because photons are waves, as well as particles, and since all particles come from photons, all particles are also waves, and thus all particles have a characteristic wavelength, which must fit into the circumference of the orbit of that particle.

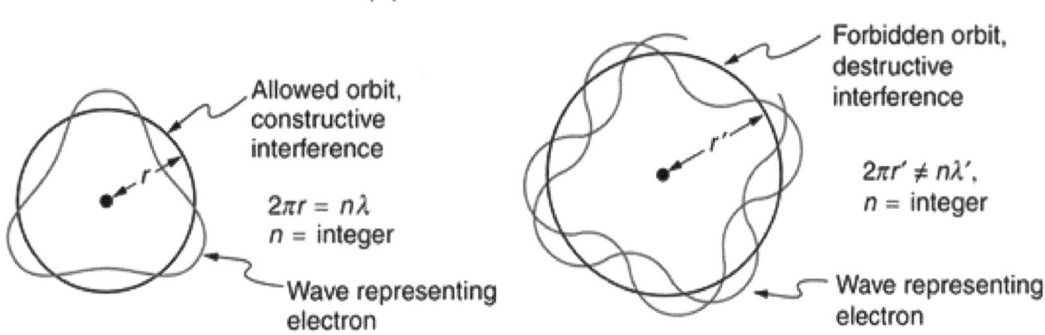

Figure 12.3. Electrons can only exist in certain orbits because they are waves as well as particles. Since celestial objects operate like superatoms, it is likely that planets also can only occupy certain orbits in a star system and that gravitational energy is quantized differently for each star system.

In conclusion, photon energy, which exists within particles, can be in two forms: heat and gravitational photon energy. In order for a photon to turn into gravitational energy, it must have certain specific wavelengths, which are characteristic of a specific atom, and depend on the position of the atom's energy levels. Photon energy not at these specific wavelengths turns into heat energy.

Reference:

[1]   Albers, C. (2018). Article 336: Stellar nebular cloud structure

Chapter 13

338. The Planet X effect: heating and ionization in contact regions

The Planet X effect refers to the heating and ionization which occurs whenever Stellar Core matter makes contact with Solar System matter. Now, in Article 336: Stellar nebular cloud structure [1] as in many previous articles, I mentioned that celestial bodies, i.e. planets and stars, operate like superatoms. In article 336, I also wrote about celestial objects' electron clouds, where the electrons in those clouds have a specific energy, which increases with distance from the object, and are thus like the energy levels of an atom. So, it is important to understand how energy absorption and emission works in atoms, in order to understand how it works with star systems or celestial objects.

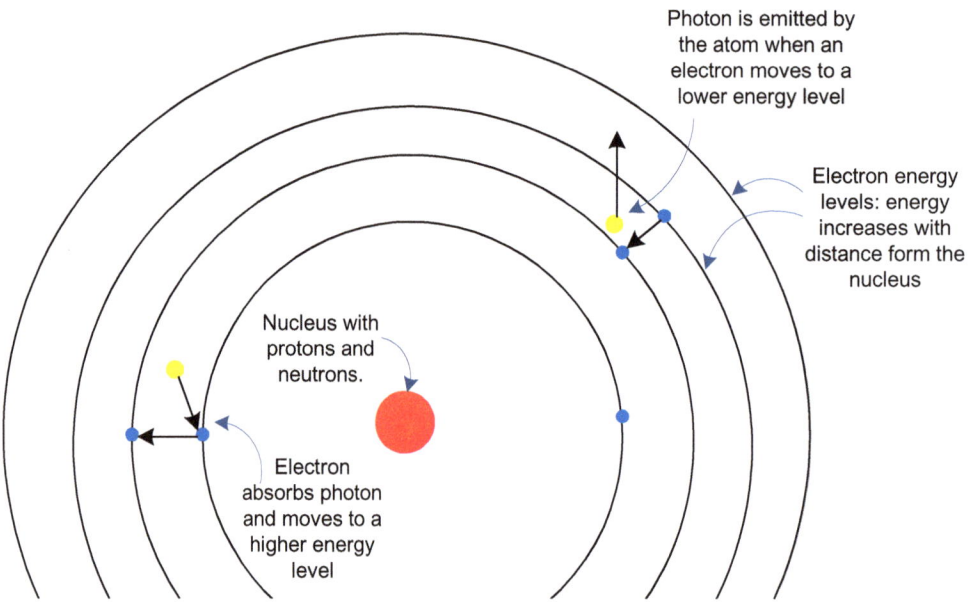

Figure 13.1. Electrons occupying energy levels further from the nucleus have higher energy. Thus, the energy associated with each energy level goes up with distance from the nucleus. Photons are emitted when electrons move to lower energy levels and photons are absorbed when electrons move to higher energy levels

Now, we know that when an atom absorbs energy in the form of a photon, it often leads to an electron leaving the atom, i.e. the atom becomes ionized, and we know that an atom emits a photon, i.e. it releases energy when an electron moves to a lower energy level. But what would induce an electron to move to a lower energy level in the first place? I, like Einstein, 'do not believe that God plays dice with the universe'. In other words, this does not happen by chance, there must be a reason why the electron is induced to move to a lower energy level.

Now, as I explained in Article 260: Planet X in the Solar System: the principle of equal energy sharing [2], particles interact through their gravitational field, and equally share their energy when they come close enough to each other. This is the reason why energy in the form of heat flows from one object to another as long as the objects differ in energy but once the energy has equalized, the energy transfer stops. This means that a particle absorbs or emits energy, as a result of the other particles in its environment. Thus, an electron moves to a lower energy level, in an atom, when the atom encounters another atom, which has less energy than it does. And conversely, an atom, which encounters other atoms, with more energy than it has, will absorb photons emitted by the atoms with more energy. This causes an electron, in the atom absorbing the photon, to move to a higher energy level. Thus, energy absorption or emission is dependent on the environment. An object with more energy, than its environment, will automatically emit energy, in the form of photons, or light. However, atoms emit specific wavelengths of energy, which are different for different types of atoms, as these are dependent on the position of each atom's energy levels. Thus oxygen atoms emit photons of different wavelengths from carbon atoms, for example. So, if two carbon atoms share energy, this energy will be absorbed as gravitational energy, but if a carbon atom and an oxygen atom share energy, then the energy is not at the right wavelength to be absorbed as gravitational energy, and thus, will be absorbed as heat, instead. But in the case of an outer electron absorbing a photon, any photon energy may result in the electron leaving the atom, which then becomes ionized (see Article 337: Heat and gravitational photon energy) [3].

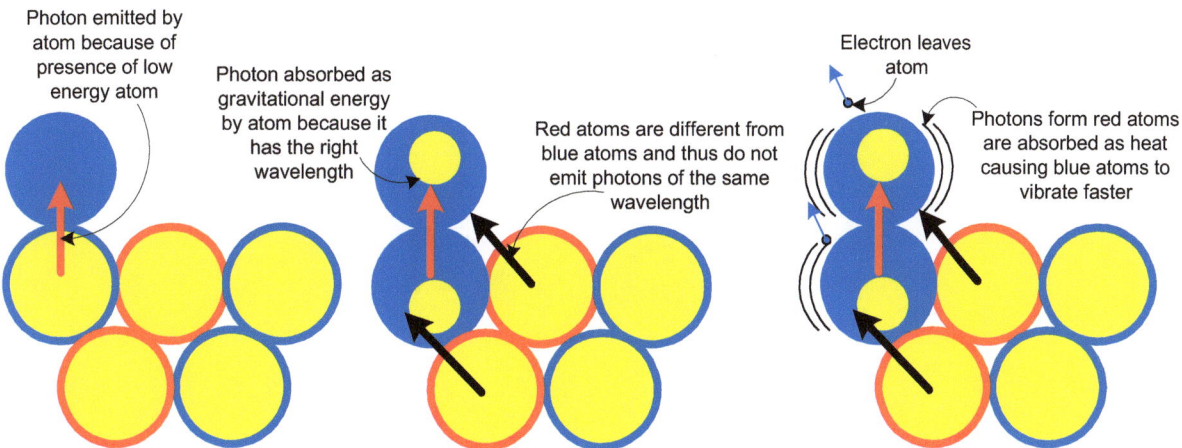

Figure 13.2. Photons emitted by atoms can be absorbed as gravitational energy, which allows electrons to move to higher energy levels, or as heat energy, which causes an atom to increase its rate of vibration and may also cause an electron to leave the atom and thus become ionized. The greatest heating and ionization will occur at the layer, where the greater amount of energy transfer occurs, which is the region where contact is made.

In addition, in the case of a celestial object, i.e. a star or a planet, the loss of energy also triggers the flow of energy from the object's core, out toward the outer layers. This also occurs as a result of the principle of equal energy sharing. Particles, lower down, inside the object, lose energy to other particles, closer to the surface, resulting in the energy transfer to spread through the whole object. It eventually reaches the core. When the particles in the core start to lose gravitational energy, this triggers an increase in energy generation by the core, which then flows outwards toward the outer layers (see Article 240: Planet X System effect on radioactive decay rate and heating of planets) [4]. This causes energy to be emitted from all levels of the object, all the way to the surface, and causes heating, and increased ionization, but the layer that will experience the most heating, and ionization, is the layer, which has come into contact with the lower energy particles. This is most likely to be the top layer unless the Stellar Core matter penetrates into the planet's atmosphere.

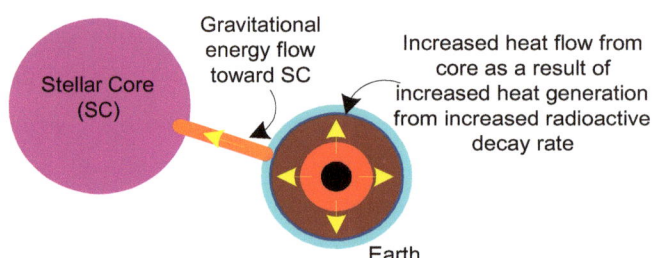

Figure 13.3. A Stellar Core's low energy status leads to it absorbing gravitational energy, from a planet, which eventually causes a decrease in the core's gravitational energy, which thus triggers the radioactive decay rate in the core to increase, and thus the core's energy generation to increase. This increases the rate of energy flow, to the surface, and thus increases the planet's surface temperature. In the case of the earth, it has greatly increased its ocean temperature (see Article 240: Planet X System effect on radioactive decay rate and heating of planets) [4].

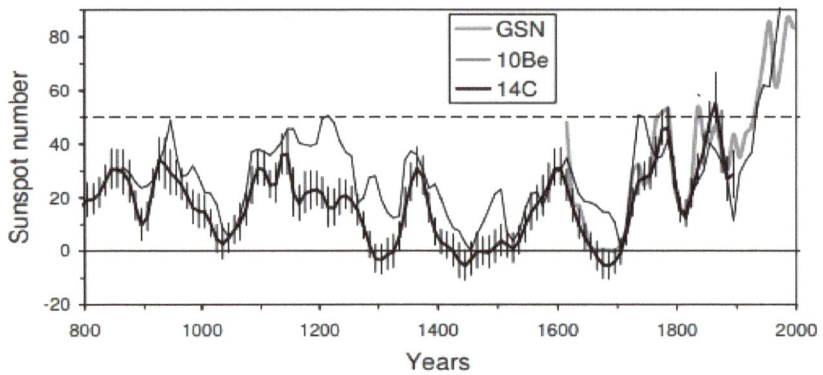

Figure 13.4. Plotted 10-year averaged sunspot numbers (thick grey line), and reconstruction coming from ice core sample data (thick grey line with vertical lines and black lines) [6]. Solar activity has never been as high, as around the year 2000, in 1200 years.

Thus, Stellar Core matter causes the heating and increased ionization of all earth's layers but especially the layer where contact is made. This is why Stellar Core dust entering the earth's upper atmosphere gives rise to noctilucent clouds, i.e. clouds that emit light in the upper atmosphere. Thus, the presence of Planet X System objects inside the Solar System leads to heating and ionization, especially in the outer layers of Solar System objects, at least for as long as they have energy reserves in their cores. The Sun seems to now be running low on energy reserves in its core and is thus going dark, but its activity greatly increased between 1850 and the year 2000 (see Article 3: Interplanetary Climate Change: Planet X and the current Grand Solar Maximum) [5]. The year 1850 is also the year that noctilucent clouds started being observed on earth and thus the year when Stellar Core matter started entering the atmosphere. In other words, this was the year when the Planet X effect was first observed.

In conclusion, the energy transfer mechanism, from earth matter to Stellar Core matter, causes heating and ionization of the earth matter, which makes contact with the Stellar Core matter. Thus, the presence of Planet X System objects and matter, such as the debris, which these objects generate, gives rise to the Planet X effect, which is the heating and ionization in any contact region, and which is likely to occur to a greater degree in the outer layers of Solar System objects (planets and stars) and the Solar System itself (at the boundary of the Solar System and interstellar space).

Reference:

[1] Albers, C. (2018). Article 336: Stellar nebular cloud structure.
[2] Albers, C. (2018). Article 260: Planet X in the Solar System: the principle of equal energy sharing.
[3] Albers, C. (2018). Article 337: Heat and gravitational photon energy.
[4] Albers, C. (2018). Article 240: Planet X System effect on radioactive decay rate and heating of planets (in Book 7: Planet X The effects on the Earth and the Sun).
[5] Albers, C. (2018). Article 3: Interplanetary Climate Change: Planet X and the current Grand Solar Maximum.
[6] Usoskin, I. (2017). A history of solar activity over millennia. A history of solar activity over millennia. Living Rev. Sol. Phys. 14:3 https://doi.org/10.1007/s41116-017-0006-9

Chapter 14

339. Planet X and the Interstellar Medium: can we leave the Solar System?

The interstellar medium is the space outside the Solar System. The Solar System is made up of the Sun's Heliosphere, a close to the spherical bubble created by the Sun's Solar Wind and magnetic field. It is the Sun's equivalent to the Earth's magnetosphere, and it is thus expected to deflect the interstellar wind away from the Sun. Hence, the heliosphere acts as a protective shield, which keeps most cosmic rays, in other words, most ionized or charged particles from entering into the Solar System. The Heliosphere was initially believed to be egg shaped and to have a long tail, behind it, but new data indicates that it is spherical and has no tail.

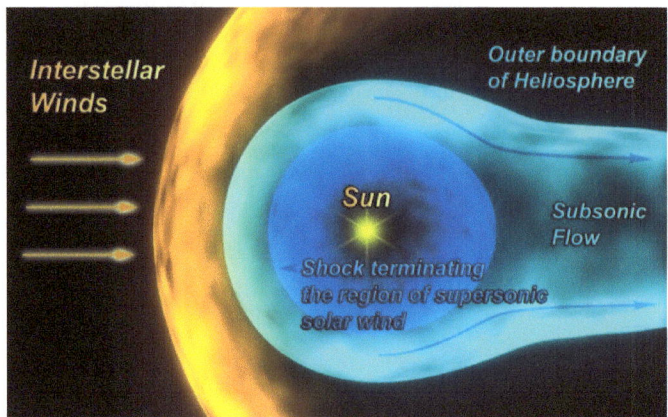

Figure 14.1. In 2008, data from the Ulysses spacecraft showed that the interstellar wind impacting the Heliosphere was fast enough to create a shockwave outside the Heliosphere [1].

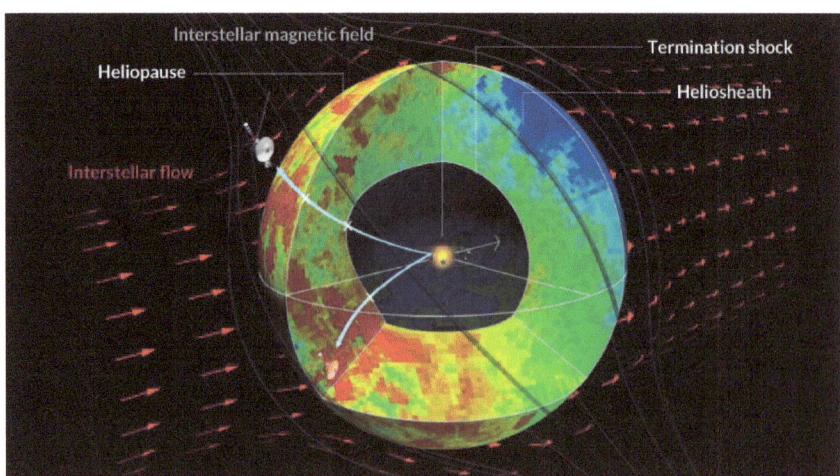

Figure 14.2. In 2017, data from the Cassini and Voyager spacecraft revealed that the Solar System did not have a long tail but was spherical [2].

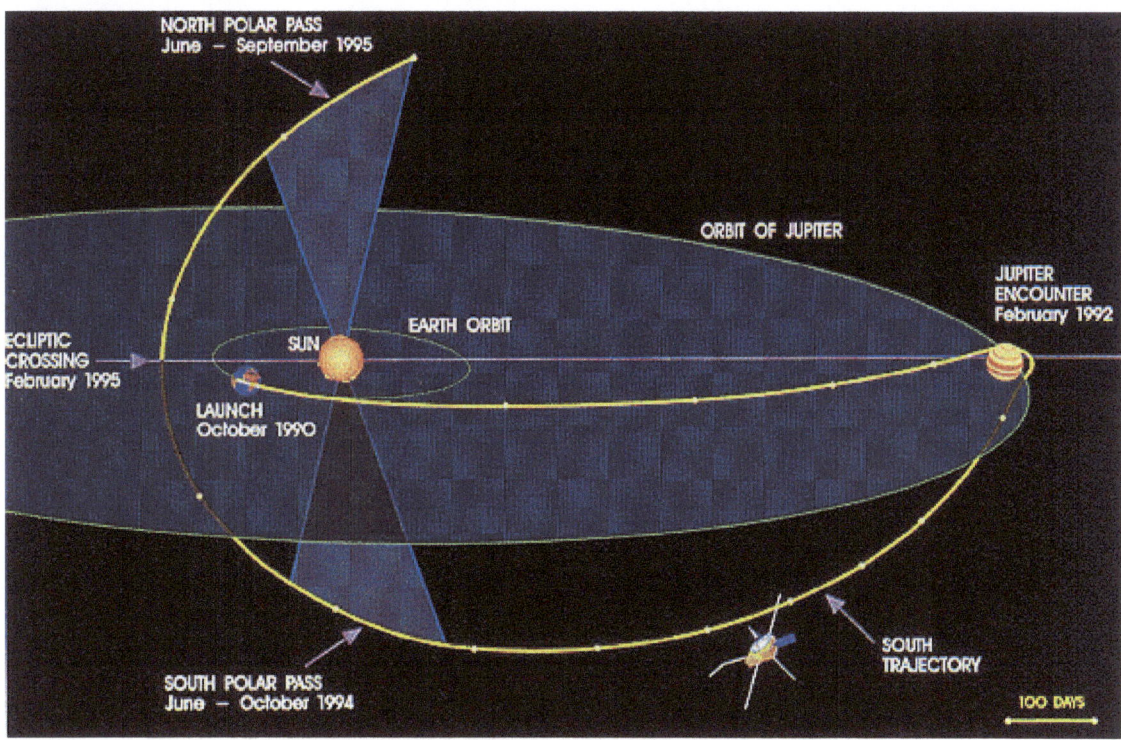

Figure 14.3. Ulysses' first polar pass after launch, in 1990.

The Ulysses spacecraft was launched in 1990, into a solar polar orbit. It was designed to study the Sun at all latitudes including the Sun's poles. The spacecraft did not orbit close to the Sun; its perihelion position was 1.35 au and it aphelion position was 5.4 au, which is just beyond Jupiter's orbit; Jupiter's average orbital distance from the Sun is 5.2 au. In fact, the spacecraft flew close to Jupiter in 1992. Ulysses was also used to study the interstellar medium by detecting interstellar dust and neutral helium that had come into the Solar System from outside. The data showed that there was 30 times more interstellar dust coming into the Solar System than expected. By 2005, Ulysses' data had indicated that the Solar System was moving through a warm tenuous cloud of interstellar dust and gas. This cloud was believed to be one of several that make up the local galactic environment, through which the Sun is moving. The temperature of this cloud was determined to be 6300 K. This is not, however, a temperature that would be measured with a thermometer; this temperature is determined by averaging the speed of particles, in the cloud of gas, and is thus a measure of the heat energy contained in the particles.

The local interstellar medium had been studied between 1987 and 2011, through optical and UV line absorption, in the light of nearby stars, and thus as far out as 78 light years. These studies showed that the interstellar medium was made up of roughly equal amounts of neutral and ionized gas, which indicated that it was warm, of low density and partially ionized. In other words, the Ulysses data seemed to agree with these studies, which indicated that the interstellar medium was warm and not hot [3].

Now, the IBEX (Interstellar Boundary Explorer) spacecraft was launched in 2008, into an earth orbit, and was designed to study the interaction between the solar wind and the interstellar wind, and therefore the boundary between the Solar System and Interstellar Space, and the data showed, by 2015, that the interstellar Solar wind was much hotter than what the Ulysses data had indicated. The IBEX data suggested a temperature of 7500 K, which would make the local interstellar cloud hot rather than warm as the Ulysses data had suggested [3].

This means that the interstellar medium either became hotter between 2005, when the Ulysses data were analyzed and a temperature calculated, and 2015 when the IBEX data was analyzed, or it may have been much hotter than predicted, in the first place. I think that since the UV measurements agreed with the Ulysses data, indicating that it was just warm, up to 2011, that the temperature of the interstellar medium at the boundary, between it and the Solar System, has very recently increased, from warm to hot. So, where would this heat come from? Well, we know that there are a huge number of Planet X System objects, in the Solar System, as large numbers have been found in the Sun's corona (see Article 321: Huge Planet X star in the inner Solar System and Article 333: Huge numbers of Planet X System objects in coronagraph images) [4, 5].

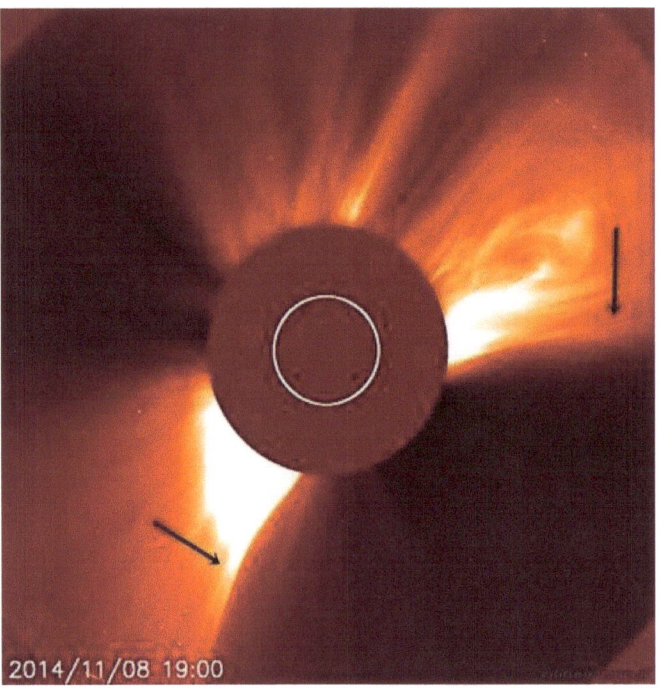

Figure 14.4. An extremely large object appears in this LASCO C2 image from 2014, sent to me by R. Wayne Steiger. The Sun appears to have a strong reaction to its presence suggesting it is very close to the Sun. This object turned out to be 7 times larger than the Sun [4].

These objects are basically dead stars and thus energy depleted. They, therefore, absorb energy from Solar System objects, which results in heating and ionization in the contact region, in what I have named the Planet X effect, as I have detailed in Article 338: The Planet X effect: heating and ionization in contact regions [6]. The planet X effect arises as a result of particles emitting photon energy, whenever they come into contact with matter, which is lower in energy. This means that heating would be expected to occur anywhere where the Planet X System of Stellar Cores makes contact with the Solar System and since the Planet X Objects must come from outside the Solar System, the boundary between the Solar System and the interstellar space, occupied by the Planet X System, is the likely region where this heating and ionization would take place. In addition, the recent increase in the temperature at the outer edge of the Solar System indicates that larger numbers of objects and their debris have approached the outer edge of the Solar System, in the last few years.

Figure 14.5. SOHO LASCO C1 image from June 25th, 1986, showing a large jet plasma ejection indicating the likely presence of Stellar Cores in the Sun's corona, discharging between each other. There are thousands of dark centers surrounded by disk shaped rings, indicating the presence of point sources of light, as viewed through a telescope which produces diffraction effects.

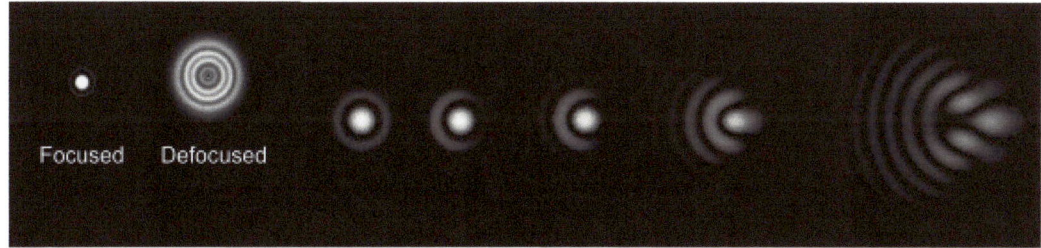

Figure 14.6. A star or other distant light source looks like alternating bright and dark concentric rings when observed through a telescope due to diffraction effects produced by the eyepiece [4]. If in addition, telescope components such as mirrors are not perfectly aligned further effects can be produced which may make a bright point source appear to be a series of concentric and alternating bright and dark arcs. When everything is perfectly aligned then the telescope is described as being in collimation.

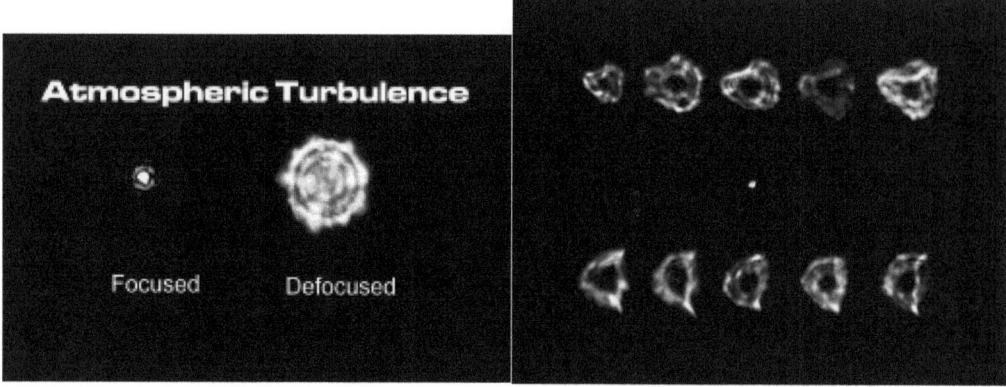

Figure 14.7. Left: Atmospheric turbulence can also affect the way a bright source of light looks like through a telescope. **Right:** Optical defects such as an uncollimated beam in a lens system in addition to atmospheric turbulence may cause light sources to appear to be made out of a dark center with distorted bright concentric rings around them.

Data provided by several spacecraft reveals that the Solar System is within a cloud of dust and gas, called the Local Interstellar Cloud and since the Solar System seems to be close to one edge of this cloud, it seems to have recently entered it [7]. Obviously, the Planet X System would have to be inside this cloud as well. As this cloud seems to be quite uniform, it is therefore possible that the Planet X System occupies the whole cloud, and that therefore the cloud is the Planet X System, but that, until recently, we had only moved through the outer edges of it, where there were not as many objects or debris, as in the part that we encountered after 2011, since that was the time when the last of the UV data indicating a warm interstellar environment was published. So it seems that between 2011 and 2015 we encountered a more densely populated part of the Planet X System, which must have caused many more, larger objects, and debris to enter the Solar System and arrive at the Sun's corona. And it also caused more Stellar Core matter to interact with the Solar System at the boundary of the Solar System, which then caused the boundary to heat up. This heat comes from the Sun; it is transmitted through the Solar System out to the edge. The heat does not come from the Planet X Objects because they do not have the energy to share, they are absorbers of energy. However, this also suggests that the Planet X Objects will not just come straight through the Solar System boundary but will remain outside at the boundary for some time, absorbing as much energy as they can, from the boundary, and only after some time will they penetrate and start making their way towards the Sun. It is also possible that comets are a part of this system since comets have to also come from outside the Solar System.

Now, the local interstellar cloud is 30 light years across and thus nearly 20 000 times larger than the Solar System. This would then suggest that the Stellar Cores that are now found in the Sun's corona are only a small percentage of the total number of objects in the Planet X System and that these objects will continue to come in as time passes. In addition, this would mean that the Solar System is surrounded on all sides by the Planet X System which would make leaving the Solar System an extremely hazardous affair. In addition, these object's extremely low gravity, as a result of their low energy status, makes it impossible to navigate the system with electrogravitic or antigravity drives. The dust and debris surrounding the objects will also make journeying out beyond the edge of the Solar System hazardous even with conventional fuel drives, which would make the journey extremely long as well. In addition, there are no living systems inside the Planet X System, they are all dead. Any living system would have had its energy depleted by the rest of the system a long time ago (see Article 244: The Planet X System: destroyer of Star Systems) [8].

Figure 14.8. A Stellar Core in the Sun's corona, about 4 times larger than earth, and thus 40% of the size of Jupiter, also has a striped appearance. The object is striped because it is an old star or a Brown Dwarf

The Voyager spacecraft were launched in 1977 and were designed to explore the outer reaches of the Solar System and beyond. The Voyager spacecraft had by 2009 reached the outer bounds of the Heliosphere and analysis of its data, in 2009, indicated that the interstellar space just outside the Solar System had an unexpectedly strong magnetic field, which was tilted by 20 to 30 degrees from the interstellar medium flow direction. The magnetic field was twice as strong as expected [7]. This correlates with the fact that the Planet X objects or Stellar Cores have high magnetic fields. These objects are all over the galaxy and are known as Brown Dwarfs. However, Brown Dwarfs are actually old or dying stars, not substellar objects (see Article 317: Planet X System: Brown and Black Dwarfs with the high magnetic field) [9]. The objects lose their gravitational fields and electric fields, as they age, but retain most of the magnetic field strength they had as living stars. This would therefore suggest that the unexpectedly high magnetic field in the local interstellar cloud is as a result of the Planet X System objects, which are mostly not visible, due to their inability to emit light, as a result of their very low electric fields, in their outer layers (see Article 184: Stellar Core evolution) [10].

In addition, the Ulysses spacecraft also detected interstellar dust for the first time in 1993. Interstellar dust had been predicted to be coming into the Solar System in 1976, but the Ulysses data showed that there was 30 times more dust than expected [11]. Since there is now a huge amount of Stellar Core debris, in the Solar System, and it was most likely the entrance of Stellar Core dust, which led to the appearance of noctilucent clouds, in 1850, it should not be surprising that this dust was found in the Solar System in 1993, or that it was in fact much more abundant than predicted (see Article 146: Planet X System: time of arrival and Article 272: Noctilucent clouds and Planet X debris in the earth's atmosphere) [12,13]. It is of course likely that the amount of Stellar Core dust coming from the Planet X System will continue to increase. Then, the fact that this dust comes from different directions [7] also indicates that the Planet X System surrounds the Solar System on all sides.

In conclusion, data from various spacecraft, namely Ulysses, Voyager, and IBEX suggest that the Planet X System has surrounded the Solar System and that the system is much larger than the Solar System. This has resulted in the region of contact, the boundary of the Solar System with space outside it, reaching higher temperatures than expected and has resulted in a huge amount of interstellar dust, which is really Stellar Core dust filling our Solar System. We seem to have entered a more densely part of this system between 2011 and 2015. In addition, the Planet X System may be as large as the whole Interstellar Cloud that the whole Solar System finds itself inside of. The cloud is 20 000 times larger than the Solar System, and since these objects and their debris would make navigating through the system extremely hazardous if not impossible, this means that leaving the Solar System is most likely impossible.

References:

[1] https://www.esa.int/spaceinimages/Images/2008/06/Interstellar_wind_hits_the_heliosphere.

[2] Dialynas, K. et al. (2017). The bubble-like shape of the heliosphere observed by Voyager and Cassini. Nature Astronomy, vol.1, Article 0115.

[3] McComas, D. et al. (2015). Local Interstellar Medium: six years of direct sampling by IBEX. The Astrophysical Journal Supplement Series 220 (2).

[4] Albers, C. (2018). Article 321: Huge Planet X star in the inner Solar System.

[5] Albers, C. (2018). Article 333: Huge numbers of Planet X System objects in coronagraph images.

[6] Albers, C. (2018). Article 338: The Planet X effect: heating and ionization in the contact region.

[7] Opher, F. et al. (2009). A strong, highly-tilted interstellar magnetic field near the Solar System. Nature 462, pp. 1036-1038.

[8] Albers, C. (2018). Article 244: The Planet X System: destroyer of Star Systems (in Book 7: Planet X The effects on the Earth and the Sun).

[9] Albers, C. (2018). Article 317: Planet X System: Brown and Black Dwarf with the high magnetic field.

[10] Albers, C. (2018). Article 184: Stellar Core evolution (in Book 3: Planet X revealed Gravity and Light).

[11] Veerle, J. et al. (2015). Sixteen years of Ulysses Interstellar dust measurements in the Solar System. III. Simulations and data unveil new insights into local interstellar dust. The Astrophysics Journal 812 (141) pp. 1 – 24.

[12] Albers, C. (2018). Article 146: Planet X System: time of arrival.

[13] Albers, C. (2018). Article 272: Noctilucent clouds and Planet X debris in the earth's atmosphere.

Chapter 15

347. Gravity wave on Venus suggests Planet X presence

In 2015, a curious bulge was detected for the first time, on Venus' atmosphere by the Japanese probe Akatsuki. The event has repeated itself many times, since, and it puzzled the astronomers who had no available answers as to what could possibly have caused such bulge. The bulge has reappeared many times and it remains in the Venus' atmosphere, for days at a time, before disappearing. The bulge, which seems to stay stationary, with respect to the surface of the planet, was interpreted as a gravity wave by a group of Japanese scientists, who also thought that it was created by Venus' very high mountain ranges. The planet's rotational period is 116 days, but its atmosphere has a rotational period of 4 days, and since the bulge stays in position, with respect to the surface, the Japanese scientists thought that it must be created by the mountain ranges over which it forms [1].

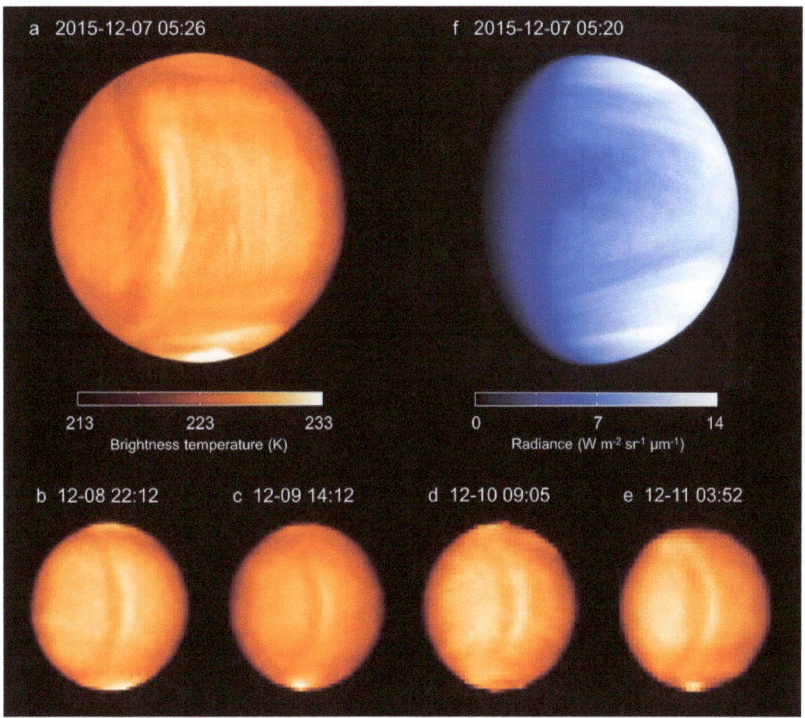

Figure 15.1. Bulge as when it was observed in the Venus atmosphere in 2015. It did not change position for 4 consecutive days even though the atmosphere would have completed one rotation in that time. The wave is in the form of an arc as if it is a part of a ring. It can be discerned on the radiance image, so its crest is both brighter and hotter than the surrounding atmosphere.

However, this had never been observed before, and Venus has been observed and studied for many years now. The orbiter of the Pioneer Venus probe, orbited Venus, between December 1978 and October 1992; one of its primary objectives was to study Venus' upper atmosphere and ionosphere. This probe never detected a bulge on Venus' atmosphere. Venus' mountain ranges are not new, so if they were the cause of such a phenomenon then that phenomenon should have been observed before.

So what could be causing this to occur on Venus? Well, the Japanese scientists were I think correct that such a bulge would indicate a gravitational wave, in other words, a gravitational disruption of Venus' gravity, which can only be induced from outside of the planet, i.e. by a massive celestial object closely approaching Venus. It is possible that the fact that the wave remains stationary was confusing for them, as normal astronomical bodies do not remain stationary, with respect to each other, and yet only a massive object very close to Venus and stationary, with respect to it, could create such a stationary gravitational wave.

But, we do have objects in the Solar System that have been observed to remain stationary inside the Sun's corona, with respect to the Sun's surface, and we have seen these objects also remain stationary, with respect to the earth. In addition, the earth has experienced gravitational anomalies, such as ocean recession events, which could only have been created by these objects closely approaching earth. These objects appear to be a part of a System, which I have named the Planet X System. This system is made up of the dead, or energy depleted, stars and planets, and appears to be much larger than our Solar System. The Planet X System is a destroyer of star systems; it absorbs all the energy contained in the objects of a star system thus turning them into dead objects, as well. When stars and planets become energy depleted they lose the ability to emit light and to interact gravitationally. They also lose their outer layers of material until the solid core is exposed, which is why I often refer to these objects as Stellar Cores (see Article 244: The Planet X System: destroyer of Star Systems) [2].

Figure 15.2: SDO image in the 171 angstrom wavelength from October 13[th,] 2017 showing a dark Stellar Core, which appears to be about half of the size of Jupiter, within the sun's inner corona, making a connection with the Sun and remaining stationary with respect to the Sun's surface. Its gravitational influence is creating a vortex of material, from deeper within the Sun, (chromosphere) through which it is drawing material towards itself.

Figure 15.3. Left: Ocean recedes leaving boats sitting on mud, in the harbor in Punta del Este, Uruguay, on August 11th, 2017. The ocean came back, showing the event was tidal in nature, but this extreme low tide had never happened before. There were no storms nearby to try and blame for this event. **Right:** People walk out onto the sand, in Tampa Bay Florida, on September 10th, 2017. The beach was left empty due to the ocean receding. A hurricane affecting the area passed overhead, before the water started returning, the next morning, showing that the ocean recession did not occur as a result of hurricane winds, offshore. It does, however, indicate that an attempt was made to cover up this tidal event with a hurricane (see Article 227: Stellar Cores affecting earth and possible connection to Volcanic Eruptions) [3], and this has again been attempted recently with Hurricane Florence (see Article 343: Hurricane Florence: cover-up for tidal force from object in outer space) [4]. However, it is likely that both the hurricane and the tidal event were created by a Stellar Core.

Figure 15.4. Image obtained from a video by the Youtube channel Jeff P. The image comes from a web camera over Germany from October 31st, 2016. Three light sources can be seen in the image. The top one is orangey pink, the middle, and brightest, is white, edged by pink light, and the lower one is white. Chemtrail clouds in front of the objects show that they are real objects in the sky.

Figure 15.5. Several more frames, from the same day, and the webcam as the image in figure 4. These frames show the Sun simulator move as the other two light sources remains stationary. The fact the two smaller bright sources remain stationary means that they are not attached to the Sun, as if they were, they would seem to move with the Sun in the sky. They must instead have made a connection with the earth, and are thus able to remain stationary with respect to the earth, in other words, they are most likely Stellar Cores inside the earth's atmosphere (see Article 243: Earth hosting at least 3 Planet X System objects) [5].

The fact that Stellar Cores, in the Sun's corona, appear to remain stationary, within the Sun's atmosphere, or corona, and those reaching earth also remain stationary, with respect to the surface of the earth, for extended periods of time, suggests that one of the objects, a Stellar Core, is responsible for the bulge, or gravitational wave, on Venus. In other words, a Stellar Core very close to Venus was responsible for the observed bulge. The fact that the same bulge has continued to be observed suggests that the same object is closely approaching Venus and causing the bulge. The object obviously remains stationary, with respect to the planet, for long periods of time, and then moves away, only to return. Stellar Cores have been observed doing the exact same thing with the Sun, for a long time now, and this shows that these objects are not just congregated around the Sun, but also around other Solar System objects. Stellar Cores are energy absorbers and any planet that is still generating energy, at its core, is thus attractive to them.

In conclusion, a Stellar Core, which has attached itself to the planet, is most likely responsible for the gravitational bulge, which has been observed appearing on Venus' atmosphere, since 2015. The continued presence of such an object, in close proximity to Venus, suggests that these objects have attached themselves to other Solar System objects, in addition to the Sun.

References:

[1]   Fukuhara, T. et al. (2017). Large stationary gravity wave in the atmosphere of Venus. *Nature Geoscience* 10, pp. 85–88.

[2]   Albers, C. (2018). Article 244: The Planet X System: destroyer of Star Systems (in Book 7: Planet X The effects on the Earth and the Sun).

[3]   Albers, C. (2018). Article 227: Stellar Cores affecting the earth and possible connection to Volcanic Eruptions (in Book 6: Planet X Physicist Articles Part 1).

[4]   Albers, C. (2018). Article 343: Hurricane Florence: a cover-up for the tidal force from an object in outer space.

[5]   Albers, C. (2018). Article 244: The Planet X System: destroyer of Star Systems (in Book 7: Planet X The effects on the Earth and the Sun).

Chapter 16

348. Venus bulge: gravitational waves and hollow planets

In Article 347: Gravity wave on Venus suggests Planet X presence [1] I detailed how a gravitational wave was discovered on Venus, in 2015, which has since reappeared many times and that since the bulge remains in place, in the same position with respect to the surface of the planet, that it has to be the result of Venus having captured or hosting a Planet X object or Stellar Core and how this suggests that these objects attach themselves to Solar System planets as well as to the Sun.

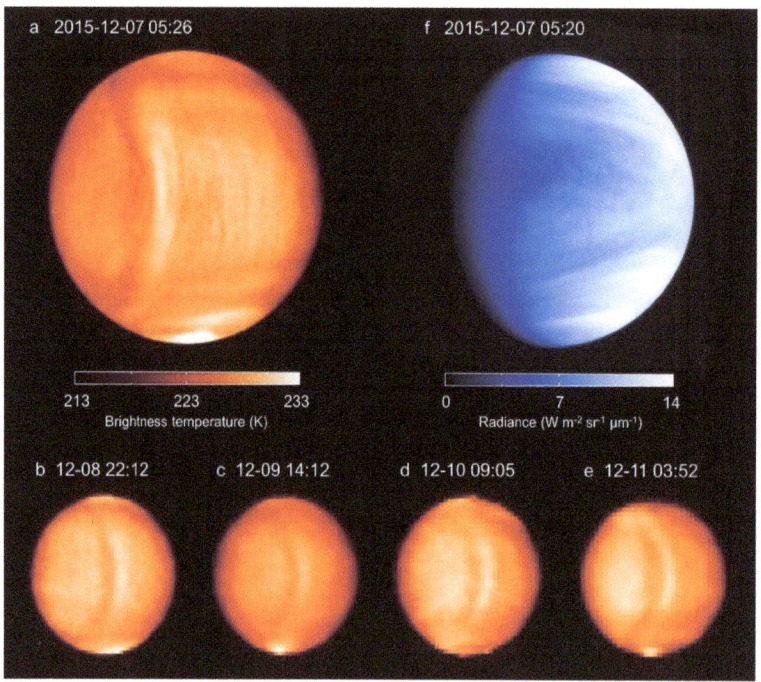

Figure 16.1. Bulge as when it was observed in the Venus atmosphere in 2015. It did not change position for 4 consecutive days even though the atmosphere would have completed one rotation in that time. The wave is in the form of an arc as if it is a part of a ring. It can be discerned on the radiance image, so its crest is both brighter and hotter than the surrounding atmosphere.

Now, there is something very interesting about the shape of the gravitational wave. It is not in the shape of a cone as I have theorized in Article 345: Hurricane Florence tidal event and gravity waves [2] and Article 346: Planet X reveals that gravity is a wave frozen in time [3]. It is, instead, in the shape of a ring, with a hollow on the inner side of the ring. This suggests that it produces a low and a high tide but that the high tide is more pronounced then the low tide. This, therefore, suggests that the gravitational wave associated with the gravitational interaction has a central minimum, not a central maximum and that the amplitude of the first maximum is higher than that of the central minimum.

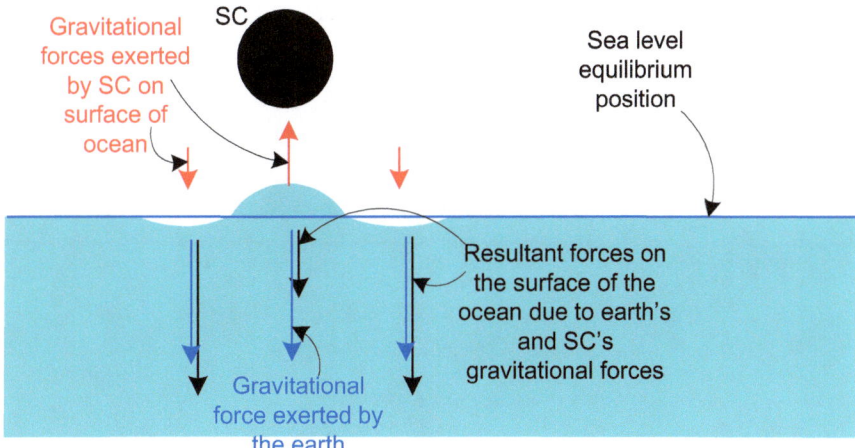

Figure 16.2. In Article 346: Planet X reveals that gravity is a wave frozen in time [3], I theorized the gravitational wave looked like a circular diffraction pattern, with a central maximum and a minimum surrounding it, of lesser amplitude than the central maximum.

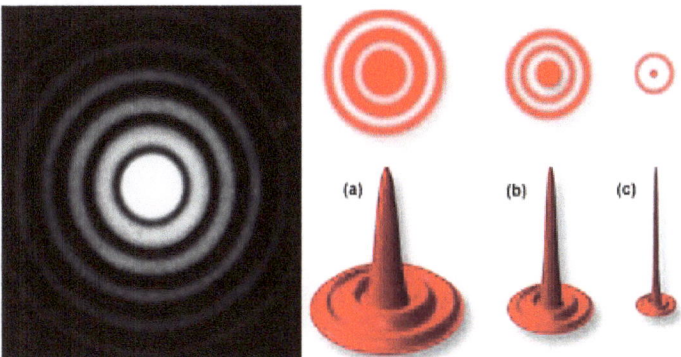

Figure 16.3. According to the wave pattern shown in figure 5, the gravitational wave would be spherically symmetric and a cross-section through it would look like a circular diffraction pattern where bright means attractive and black means repulsive. An object approaching earth would thus create a wave pattern with a central spike.

But the wave pattern seen on Venus is not like that. It is wider along the vertical direction and is curved as if it is an outside ring, not a central maximum, on a circular diffraction pattern. Also, the depth of the trough in the inner part of the ring seems less than on the outside but its depth does not seem to be as high as the height of the crest of the wave. This indicates that the stationary gravitational wave pattern is more like what is shown below:

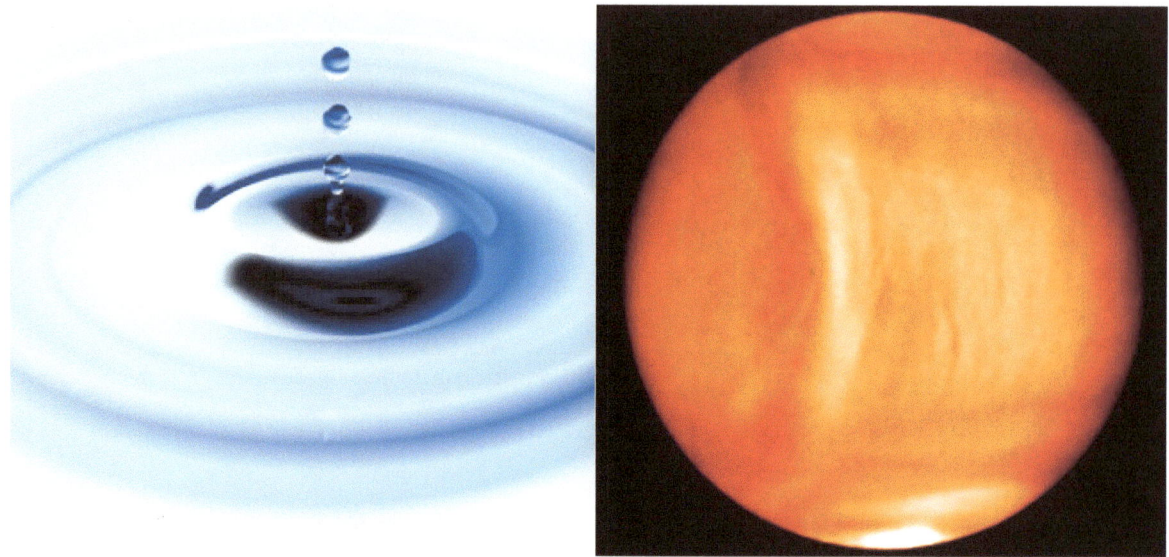

Figure 16.4. Left: A water wave with a central minimum of lower amplitude than the 1st maximum and lower amplitude than the 1st minimum, generated by a drop of water falling on a water surface. **Right**; The stationary gravitational wave on Venus: It seems to carry on to the back of the planet. The shape of this wave suggests that a gravitational wave will have a central minimum but its amplitude will be less than that of the first maximum and the first minimum.

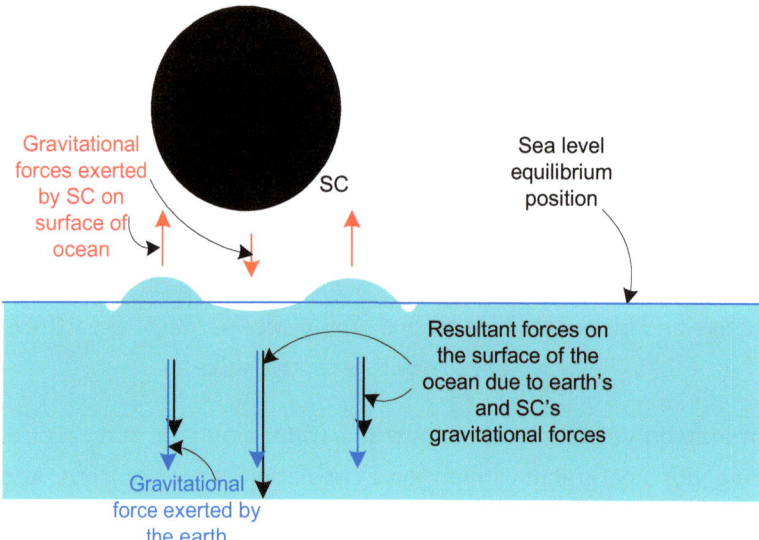

Figure 16.5. The stationary gravitational wave on Venus suggests that the gravitational wave pattern created by a Stellar Core close to the earth's surface has a central minimum, not a central maximum.

The object creating the wave pattern, on Venus, was, most likely, a little distance from the planet and on the left, or night side, of the planet so that a line running through the center of the object would pass through the central minimum of the stationary circular wave pattern, as shown below.

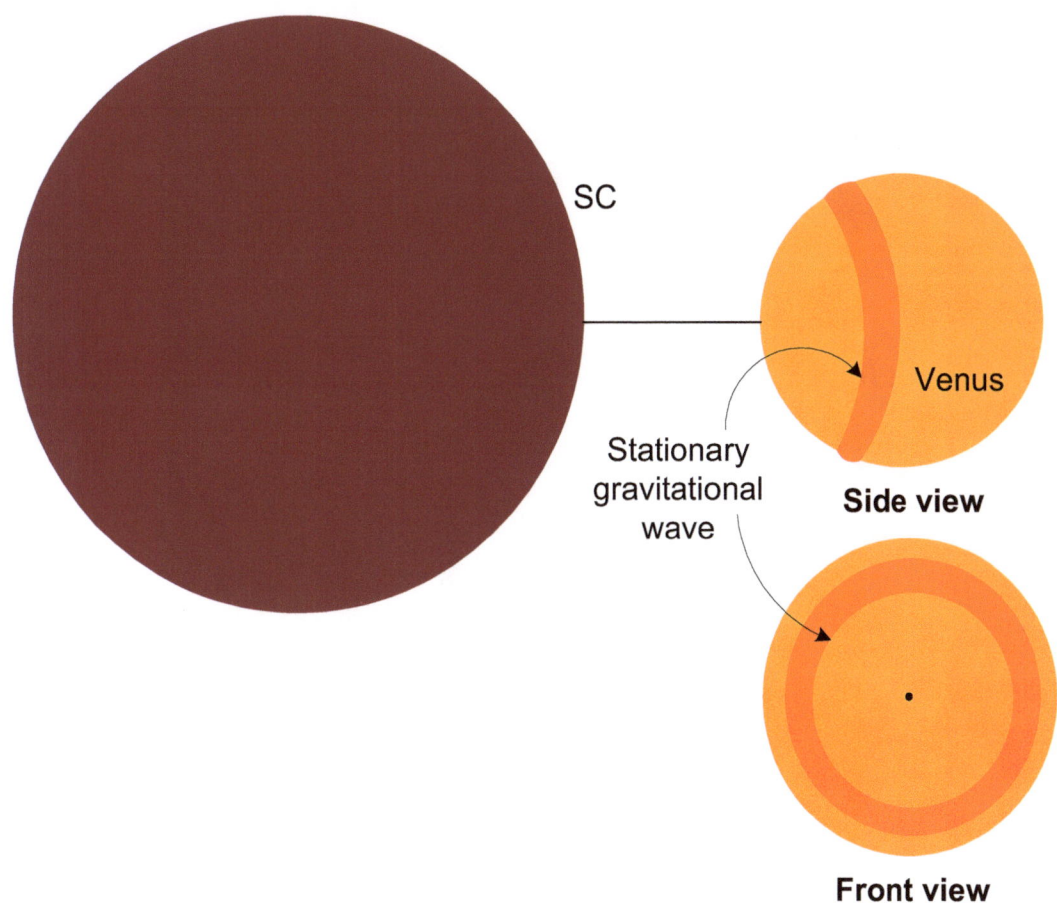

Figure 16.6. The position of the Stellar Core giving rise to the stationary gravitational wave on the planet Venus.

The radius of the gravitational wave ring will decrease as the Stellar Core, approaches the planet, and will eventually turn into a vortex.

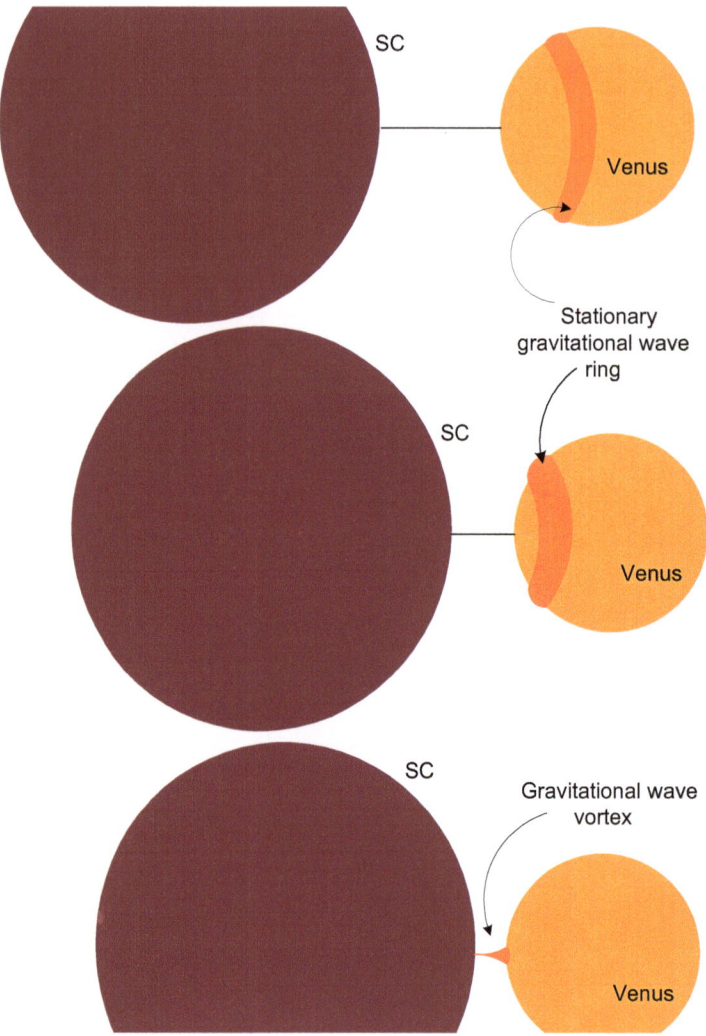

Figure 16.7. The height of the wave increases, as the Stellar Core approaches the planet, at the same time that the radius of the ring decreases, until a vortex forms between the Stellar Core and the planet.

The central minimum explains why vortices always have an empty central space. The minimum in the center is all that is necessary to cause water to spiral around the central space and it will thus occur without any need for a magnetic field. In fact, any detectable magnetic field, may, in fact, be due to the motion of the particles, as a result of the symmetry of the gravitational wave.

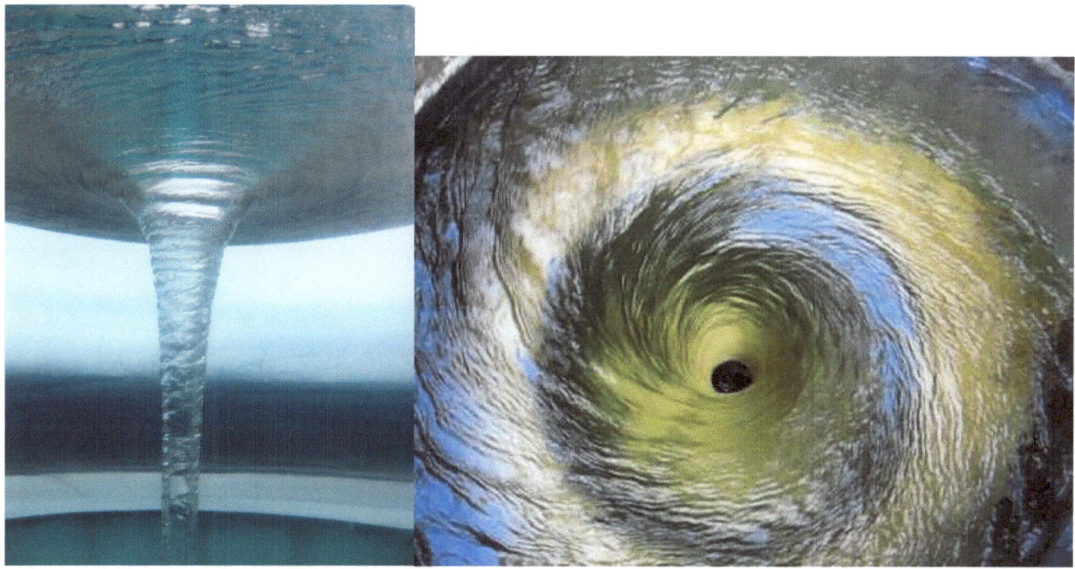

Figure 16.8. A hole at the bottom of a container, filled with water, will cause the earth's gravitational attraction exerted through that hole, on the water, to create a vortex of water.

The fact that gravity is a stationary wave, or a wave frozen in time, with a central minimum means that all massive celestial objects must have a central hollow space. Thus, all cores of planets, or stars, will have a central spherical hollow space, in them. This is because any matter moving into this region will be repelled, from that region, and thus move away toward the region where gravity is again attractive.

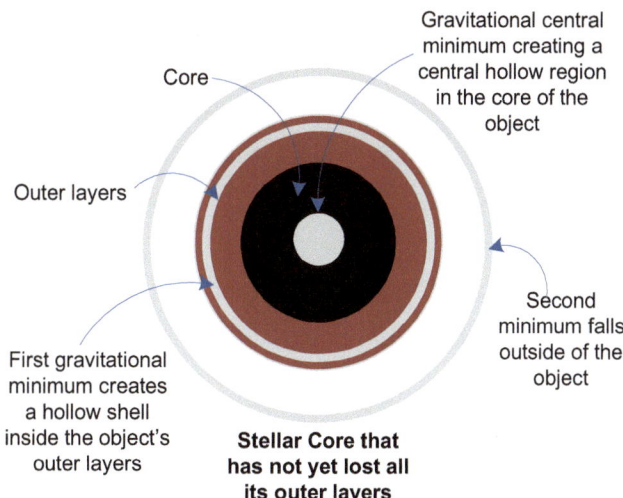

Figure 16.9. A Stellar Core that has not yet lost all its layers: It has a central hollow region in its core, due to the nature of the gravitational wave. It most likely has a first minimum, somewhere inside it, which would create a hollow shell inside its outer layers. A second minimum may be outside of the object.

The minimums are positions where gravity reverses, but the reversals are never as strong as the regions of attraction, so that hollows, on the surface of the ocean, created by a tidal force, or the gravitational influence of a massive celestial object close to the surface of the earth, are not as deep as the bulges.

In conclusion, gravity appears to be a spherically symmetric, wave frozen in time, or stationary, with a central minimum. This suggests that all celestial objects with cores will have a central spherical hollow space, inside them. This also explains why vortices have a hollow space in the center. This is the same reason why hurricanes and tornadoes have eyes, where there is no wind circulation; these all appear to be gravitational vortices.

References:

[1] Albers, C. (2018). Article 347: Gravity wave on Venus suggests Planet X presence.
[2] Albers, C. (2018). Article 345: Hurricane Florence tidal event and gravity waves.
[3] Albers, C. (2018). Article 346: Planet X reveals that gravity is a wave frozen in time.

Chapter 17

335. Biological and Ecological weapons in use against us

Dr. Claudia Albers, Planet X Physicist

Scott C'one recently comes across, and shared with me, a document available on the US Department of Defense website from 1997. It is a transcript of a meeting between the then Secretary of Defense William S. Cohen and other people. One person, named in the transcript, who appears to have been at the meeting, is Senator Nunn. Towards the bottom of the transcript, a few paragraphs appear which mention biological and eco-weapons.

Figure 17.1. Screenshot of the top portion of the transcript from 1997, which is available at http://archive.defense.gov/Transcripts/Transcript.aspx?TranscriptID=674 [1]

This is what appears lower down:

There are some reports, for example, that some countries have been trying to construct something like an Ebola Virus, and that would be a very dangerous phenomenon, to say the least. Alvin Toeffler has written about this in terms of some scientists in their laboratories trying to devise certain types of pathogens that would be ethnic specific so that they could just eliminate certain ethnic groups and races; and others are designing some sort of engineering, some sort of insects that can destroy specific crops. Others are engaging even in an eco- type of terrorism whereby they can alter the climate, set off earthquakes, volcanoes remotely through the use of electromagnetic waves.
So there are plenty of ingenious minds out there that are at work finding ways in which they can wreak terror upon other nations. It's real, and that's the reason why we have to intensify our efforts, and that's why this is so important.
http://archive.defense.gov/Transcripts/Transcript.aspx?TranscriptID=674 [1]

In other words, there are scientists working on these things so we have to make sure that we are on top of this and we must get the answers first, we must make sure we have these weapons at our disposal first. Or is it that these scientists are not necessarily in other countries and are working for the US or simply for the shadow government of the world. After all, which is the only country that has an unlimited black budget to develop new weapons? This was in 1997, 11 years ago. If they knew so much about the possibilities then, which may not even have been possibilities but simply in the beginning stages of being deployed for use, what can they do now? This paragraph suggests that the capability for controlling the climate, creating an Ebola virus, destroying specific crops, creating earthquakes and even setting off volcanic eruptions at least within the realm of possibility in 1997 and that they were keen to develop it fully. These weapons may have been in the testing stage then. Would these weapons be in use now then? And these are very powerful weapons which can be used without anybody knowing that they are being used. If you can attack a nation, such as for example Japan, with an earthquake weapon, who can really know if it was an attack at all, since it could have been a naturally produced earthquake? What about destroying a country's ability to feed its people by destroying its crops? Who is to know that it was a weapon and not a natural disaster? Would these be used as terror weapons? We know that terrorists like to create terror or fear in their victims so that they are afraid of them but if these weapons are used, who is to know that it wasn't a natural disaster, so how would the target population be induced to fear the terrorists of these types of weapons are used by them? These do not seem to be a terrorists' weapons but weapons that countries may want to use against each other.

However, it seems that it is not so many countries that are under attack, at least not the leaders of those countries, or the military of those countries, but the people, the civilians, the majority of the earth's population is under attack. And it is under attack on several fronts. The earth's population is under attack with the chemtrail spraying, severe artificially created weather, and fires that devastate whole regions, and are started with plasma weapons. Although, the weather weapons may be used before hand, in order to create drought, in the region, first, so that the fires can be disguised as natural occurrences.

In Article 33: Artificial weather [2], I wrote that I reached the conclusion that the weather was being artificially controlled after studying the behavior of Hurricane Harvey, which devastated parts of Texas at the end of August of 2017. The hurricane moved inland and then backed up thus demonstrating that it was under intelligent control. I then found a website: www.weatherwar101.com [3], where I obtained very clear information about how extensive the manipulation of our weather has become and for how long it has been going on.

Figure 17.2. The burning of fossil fuels leads to the creation of CCNs (cloud condensation nuclei), which are too small for the production of raindrops. CCN's are small particles around which water vapor condenses into droplets of water thus creating clouds. Pollution crated by fossil fuels produces droplets which are too small to ever fall as rain. A report on the research done by Daniel Rosenfeld on this appeared on a BBC news webpage, on March 10th, 2000. The report was entitled 'air pollution stops rain'.

Thus, normal evaporation from the ocean and lakes is not enough to produce rain, so huge amounts of extra steam, which places a huge deal more of water vapor, in the atmosphere, becomes necessary.

Figure 17.3. Steam coming out of the cooling towers of power generation plants: The amount of steam coming from these cooling towers is controlled and programmed to create weather systems or clouds.

This manipulation has been going on for some 100 years, but now, they have not only learnt how to produce weather systems, but they have also learnt how to produce hurricanes and tornadoes and steer them using lasers on satellites. Once clouds are produced these can be turned into tight spiraling weather systems and therefore into hurricanes and tornadoes by creating the right conditions in the earth's ionosphere, i.e. with electrical currents in the ionosphere. Weather systems are electrical in nature and it is electrical currents in the ionosphere which cause water molecules to rise, and thus create low pressure systems. High pressure systems can also be created with a current spiraling in the opposite direction. Currents in the ionosphere can be created by ionizing specific areas of the ionosphere using microwave radiation or lasers and the weather systems that are created can then be steered by the laser satellites operating from orbit. So the whole system can be controlled from space satellites (see Video entitled: Prof James McCanney on Weather Modification https://www.youtube.com/watch?v=ewzNMUALENo) [4].

Even though severe storms, including hurricanes, tornadoes and water spouts can be produced by the gravitational interaction produced when Stellar Cores closely approach earth, these storms can also be produced through electromagnetic fields. It is also obvious that the weather has been weaponized, and is now being used against the population of the US, and other countries, such as the UK, Japan, and many East Asian countries, which have been attacked by severe typhoons over and over again. At the moment (September 6th, 2018) there are 4 tropical storms affecting the world's weather:

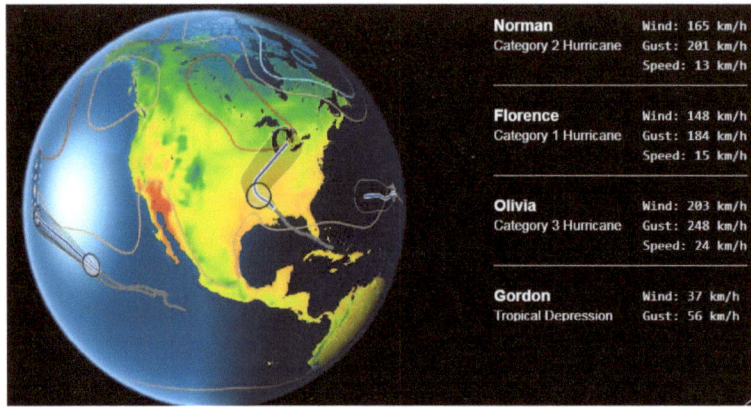

Figure 17.4. Norman and Olivia are in the Pacific, Norman is close to Hawaii, Gordon is over the US and Florence is in the Atlantic. These severe storms are used to create flooding in various areas, which destroys lives, displaces people out of their homes and creates economic hardship for the people of the region.

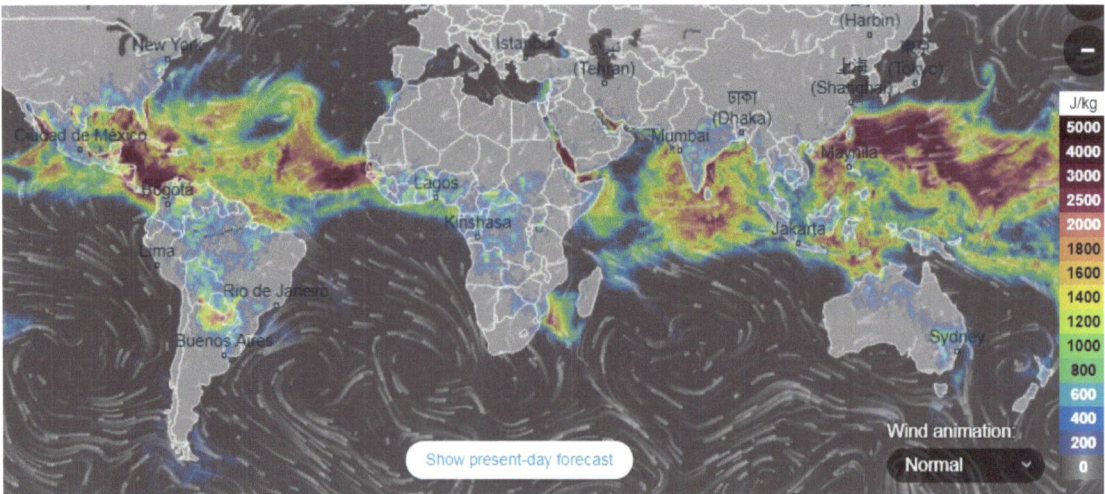

Figure 17.5. Thunderstorm activity along the equatorial regions: These storms can be created by currents in the ionosphere, which can be artificially induced. But no thunderstorms accompanied by rain would be possible without artificially creating steam.

Figure 17.6. These steam producing platforms can be placed wherever the creation of clouds is desirable and may have been placed off the coast of Africa and along the Atlantic in order to facilitate the creation of hurricanes, which can then be steered to several coastal cities in the US, and even, the UK, or Ireland.

There is evidence that earthquakes are electrical in nature. This evidence comes from strange lights which sometimes appear before and during large earthquakes (see Article 56: The Planet X system and volcanoes reveal that the universe is electrical) [5]. This means that earthquakes can be induced by currents in the ionosphere which will then induce currents inside the earth. And it also means that earthquakes can be induced through artificial induction of currents in the ionosphere and directly into the ground. Then, since volcanic activity is associated to seismic activity, as long as you target areas where there are magma reservoirs, close to the surface, it should not be too difficult to induce a volcanic eruption using the same electrical current induction process that will also provoke an earthquake or the production of a severe storm.

Thus, earthquakes, volcanic eruptions, and storms can be artificially induced through direct application of electromagnetic waves or they can occur as a result of the gravitational influence of a massive celestial object closely approaching earth. The earthquakes and storms produced through the gravitational influence of celestial objects can, however, reach a degree of severity, which artificially induced events cannot.

Figure 17.7. Volcanic lightning is an indication that rock is capable of producing lightning underground and that the earth is electrical in nature. Electrical

The large scale spraying of aerosols or chemtrails, in the atmosphere, seems to have started at the end of 1997, or beginning of 1998 (see Article 219: Chemtrails: Project Cloverleaf) [6]. Dr. Edward Teller outlined the plan to scatter millions of tons of electrically conductive metallic particulates in the stratosphere apparently as a way to reduce global warming, in 1997 [7]. This was an outright lie as these particles actually increase the greenhouse effect as detailed by a patent which is shown below. The US Air Force document entitled 'Weather as a Force Multiplier: Owning the weather in 2025' (2025 is most likely misdirection as they seem to be doing this right now) lists the many ways in which weather and climate modification can be used as weapons. Among the list is:

- Storm creation and modification
- Fog and cloud creation
- Precipitation (rain) enhancement
- Precipitation denial (creating drought)
- The artificial creation of space weather

Thus, artificial aurora or the simulation of a CME impacting the magnetosphere can be done. The chemtrail operation is carried out all over the world but the United States seems to have been the place from where it was done first. It seems to have spread to other countries after that. I did not actually see any chemtrails in South Africa until 2016. This indicates that the US military is deeply involved in this operation and that they did just what was discussed in the transcript shown in figure 1. They made sure that they had the capability before anyone else. However, since this is now being done all over the world, it means that all nations have agreed to it because if they hadn't why would not the air force of different nations shoot down these military aircraft that are spraying in these nation's airspace? The fact that the shooting down of chemtrail spraying aircraft is unheard of means that all nations have joined together, in order to fight and destroy, and deceive their own populations. This becomes very clear when one realizes that manufactured viruses and desiccated blood cells are included in the chemtrail aerosols, in addition to, heavy metals and radioactive compounds, such a thorium, which are all highly toxic to living organisms. This, therefore, implies that chemtrails are being used as a bio-weapon delivery system (see Book: Chemtrails: The silent killer) [8].

Now, chemtrails, besides facilitating the control of the climate, and the delivery of bio-weapons, also produce a greenhouse effect, which helps the atmosphere retain heat when not being illuminated by the Sun (see Article 223: Chemtrails increase greenhouse effect indicating a weakening Sun) [9].

United States Patent [19]
Chang et al.

[11] Patent Number: 5,003,186
[45] Date of Patent: Mar. 26, 1991

[54] STRATOSPHERIC WELSBACH SEEDING FOR REDUCTION OF GLOBAL WARMING

[75] Inventors: **David B. Chang**, Tustin; **I-Fu Shih**, Los Alamitos, both of Calif.

[73] Assignee: **Hughes Aircraft Company**, Los Angeles, Calif.

[21] Appl. No.: 513,145

[22] Filed: Apr. 23, 1990

[51] Int. Cl.^{5} G21K 1/00
[52] U.S. Cl. .... 250/505.1; 250/504 R; 250/503.1; 244/158 R
[58] Field of Search .... 250/505.1, 504 R, 503.1, 250/493.1; 244/136, 158 R

[56] **References Cited**
U.S. PATENT DOCUMENTS
3,222,675 12/1965 Schwartz 244/158
4,755,673 7/1988 Pollack et al. 250/330

Primary Examiner—Jack I. Berman
Attorney, Agent, or Firm—Michael W. Sales; Wanda Denson-Low

[57] **ABSTRACT**

A method is described for reducing atmospheric or global warming resulting from the presence of heat-trapping gases in the atmosphere, i.e., from the greenhouse effect. Such gases are relatively transparent to sunshine, but absorb strongly the long-wavelength infrared radiation released by the earth. The method incudes the step of seeding the layer of heat-trapping gases in the atmosphere with particles of materials characterized by wavelength-dependent emissivity. Such materials include Welsbach materials and the oxides of metals which have high emissivity (and thus low reflectivities) in the visible and 8-12 micron infrared wavelength regions.

18 Claims, 2 Drawing Sheets

Figure 17.8. US patent from 1991 describing method of using metal oxide particles of very low reflectivity (dark) for converting blackbody radiation emitted by earth to wavelengths that will be reemitted into space. It also describes the mechanism by which high reflectivity (shiny) particles increase the greenhouse effect. Chemtrail particulates have high reflectivity and so increase the greenhouse effect [8].

In addition, chemtrails because they contain huge amounts of metallic particles increase the conductivity of the atmosphere, which then facilitates the use of plasma or laser weapons. These weapons seem to also not have been designed to protect the civilian population of any country but rather to attack them, as these seem to be used to start fires (see Article 106: California population attacked with fires) [9].

Figure 17.9. A laser beam from the sky is seen here starting a fire.

These weapons can be used to target certain materials and thus explain why steel may burn whilst wood does not. The reason is that the light beam can be produced at different frequencies, which are resonant frequencies for different materials. When a material is targeted by light at the correct resonant frequency, for that material, the light superheats the atoms and molecules in that material. In other words, the atoms and molecules start vibrating faster and faster to the point that the material ignites. Thus, like more energy, is driven into it, by the beam continuing to deliver energy into it, the material reaches such high temperatures that it vaporizes or explodes and then burns, from the inside outwards as the inside reaches higher temperatures than the outside. This is a case of light energy being transformed into heat or kinetic energy of the particles making up the material.

Figure 17.10. Directed Energy Weapons (DEWs) which generate high energy laser beams, at various frequencies, mounted on aircraft. They are also mounted on satellites.

In conclusion, there is a war going on. This war is not between nations. This is a war on the civilian populations of this world. The leaders of this world have decided to mark us for extermination and they have aimed their weapons at us. The question we should, therefore, ask ourselves is: Since it seems that the Planet X System is set to destroy the surface of this planet and make life, here on earth, impossible (see Article 244: The Planet X System: destroyer of Star Systems) [10], why would they want to kill off most of the world's population now. If we are all going to die anyway why do they want to bring that end about a little sooner?

References:

[1] http://archive.defense.gov/Transcripts/Transcript.aspx?TranscriptID=674
[2] Albers, C. (2018). Article 33: Artificial weather (in Book 5: Chemtrails: the Silent killer)
[3] www.weatherwar101.com

[4] Video entitled: Prof James McCanney on Weather Modification
https://www.youtube.com/watch?v=ewzNMUALENo
[5] Albers, C. (2018). Article 56: The Planet X system and volcanoes reveal that the universe is electrical.
[6] Albers, C. (2018). Article 219: Chemtrails: Project Cloverleaf (in Book 5: Chemtrails: the Silent killer).
[7] Teller, E. and Lowell, W. (1997). Global Warming and Ice Ages: Prospects for Physics-Based Modulation of Global Change, prepared for invited presentation at the International Seminar On Planetary Emergencies, Erice, Italy, August 20-23, 1997.
[8] Albers, C. and C'one, S. (2018). Book 5: Chemtrails: The silent killer.
[9] Albers, C. (2018). Article 106: California population attacked with fires.
[10] Albers, C. (2018). Article 244: The Planet X System: destroyer of Star Systems (in Book 7: Planet X The effects on the Earth and the Sun).

Chapter 18

342. The enemy: Why are bio and eco-weapons being used against us?

In Article 335: Biological and Ecological weapons in use against us [1], I detailed the many bio and eco-weapons that the earth's population is being attacked with. These could also be referred to as deceptive weapons. They are deceptive because they are designed to seem like what is occurring is natural, as these weapons can be used to create hurricanes and tornadoes, as well as steer them. The weather has been mostly artificial, for over 100 years, but more recently it has been weaponized and almost the whole planet is now being targeted. Too much rain or drought can be artificially induced. But that is not all, Ebola type viruses, viruses which target certain ethnic groups, insects which target specific crops, and earthquakes and volcanic eruptions, induced with electromagnetic waves, are all possible, and detailed in a transcript of a meeting, in 1997, between the then Secretary of Defense William S. Cohen and other people. The transcript is available on the US Department of Defense website, and the description of the weapons appears close to the bottom.

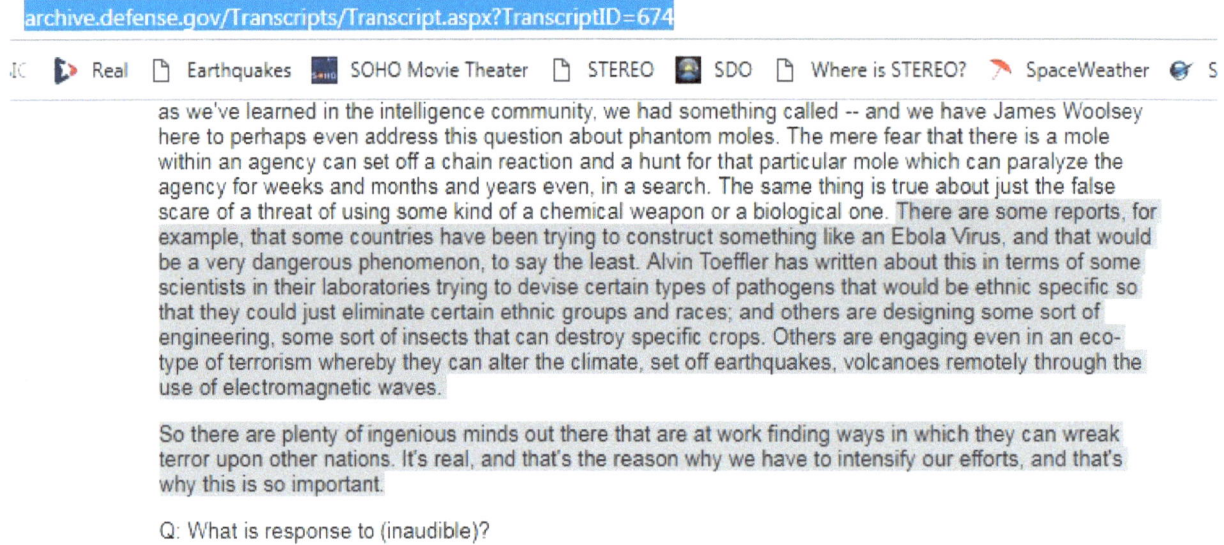

Figure 18.1. Screenshot of the lower portion of the transcript from 1997, which is available at http://archive.defense.gov/Transcripts/Transcript.aspx?TranscriptID=674 [2]

Although the weapons are described as being used by terrorists, in the transcript, it should be clear that these are not the type of weapon that a terrorist would use, precisely because their use can be mistaken for a natural event. Terrorists like to make their victims terrified of them, and if their victims think these are natural events, why would they be afraid of the terrorists?

I ended Article 335: Biological and Ecological weapons in use against us [1] with the following statement: *Since it seems that the Planet X System is set to destroy the surface of this planet and make life, here on earth, impossible (see Article 244: The Planet X System: destroyer of Star Systems) [3], why would they want to kill off most of the world's population now?*

If we are all going to die anyway, why do they want to bring that end about a little sooner? And notice that is not just the use of these weapons that are threatening our lives. It has become clear that they are posturing toward third world war from what is occurring in Syria. Russia accuses and threatens the US of harboring terrorists and threatens their positions and the US quickly sends an aircraft carrier. It was so quickly done that it almost seems that it was done with glee as if they are simply so keen to get started with the shooting. Once that starts the stage has already been set, as Russia has older equipment than the US, and they have already said that they would use nuclear weapons if they were losing a conventional weapon war. War has always been the time when most people die; it is like a culling of the human race. The third world war has the potential to kill billions of people. But why are they so desperate to have it? If Planet X is set to destroy the surface of the planet and thus destroy all life on its surface within the next few years, why is there this desire to kill billions of people a few years before they would die anyway?

Before I attempt to answer that question, I would like to consider another question: Why does our physics and astronomy teach and promote ideas that are the exact opposite of how the universe seems to operate? For example, black holes play a big part of what astronomy believes about the universe, because everything is about gravitational collapse, but the Stellar Cores, or dead stars, found in the Sun's corona in large numbers, have demonstrated that there is no gravitational collapse, that dead stars have just about no gravitational influence, and that therefore there is no such thing as a black hole (see Article 210: Stellar Core gravity: tidal and G is not constant) [4]. It turns out that matter is created by the nuclei of galaxies, so instead of a black hole, at the center of galaxies, there is what is better described as a white hole, which is the exact opposite of a black hole (see Article 288: Image of center of the Milky Way Galaxy reveals how the universe works) [5].

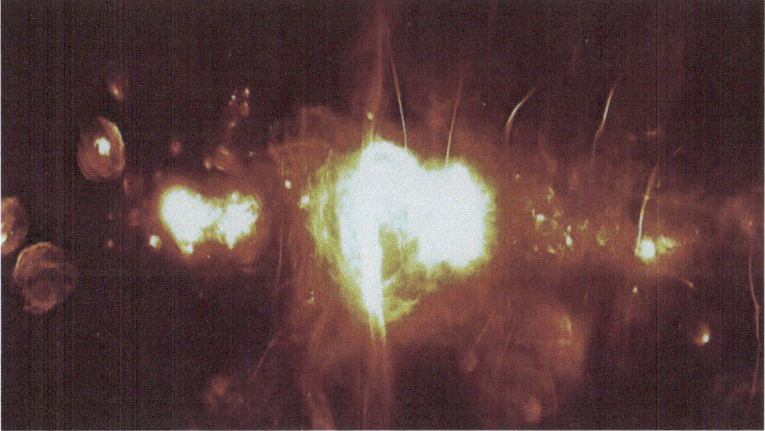

Figure 18.2. The nucleus is most likely somewhere inside the brightest part. The filaments indicate that ejection of matter is taking place from different spots. The ejections most likely condense into globular clusters, which then fission into individual stars and the nucleus itself seems to have split into pieces. This agrees with the fact that gravity decreases with age leading to expansion and fissioning [5].

We are also taught that the Sun is powered by thermonuclear reactions, which would make it impossible for the Sun to go dark, as there would be light continuously coming up from the interior of the Sun. But yet, the Sun is clearly going dark as I have shown in Article 324: Planet X causes the sun to be darkened [6]:

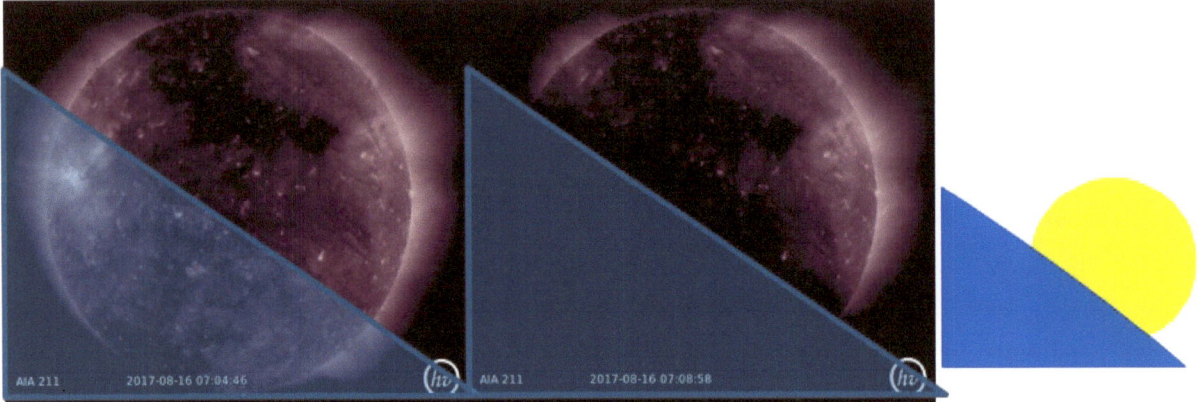

Figure 18.3. When we compare the image on left, from the first day of the second eclipse season of 2017, with one from 4 minutes after, we see that, in the later image, the corona is not covered, but it has shrunk. The earth is not curved enough to produce the darkness seen above the triangle. And, how can several cellulars like parts of the Sun remain visible below the edge of the darkness, if the earth is supposed to be covering them? The diagram on the right indicates the curvature that the earth would have if the earth was eclipsing the sun from SDO's perspective.

So the question we should ask ourselves is: why are we lied to in this manner, why are we taught the exact opposite of how the universe really works? And it isn't just in the sciences that we are taught lies. Lies are being promoted in just about every area of life. Are we given pharmaceutical drugs in order to treat disease? But how many of us realize that these drugs may make us feel better initially, but because they are not natural, they interfere with the natural balance in our body, and thus lead to more disease? Are we told that supplements are helpful to our health? But how many people realize that most supplements contain fillers, such as microcrystalline cellulose, which builds up in the digestive system, and over time blocks the absorption of nutrients, even from food? So we take supplements to increase our nutrient intake, and we end up over time absorbing less and fewer nutrients? Our forefathers ate diets based on grains and vegetables, with just small amounts of meat, except for areas with very cold climates such as Arctic regions, but we are told that we need meat to be healthy and thus in the last 50 years diets worldwide have become more and more based on meat, rather than on grains and vegetables. Is it possible that we have deviated from the natural diet, our bodies are designed for, and what do you think is the outcome of such a deviation? Is it possible that the medical and pharmaceutical industries, on this planet, are greatly benefitting from this deviation?

How many people believe that facial creams are necessary in order for a person to look their best? But how many people realize that all those small amounts of preservatives and other substances, in these creams, build up in the skin, and over time block the skin from getting nutrients, causing it to dry out and age. So the creams that we believe are going to keep us looking young, are actually doing the opposite? Why is the world's economy built on debt instead of value? It should be that as a country builds infrastructure and factories that its capital increases; that it becomes richer, but instead a country's value has been tied to how much money it can borrow, and it can get its citizens to borrow. This can never logically work. If you borrow $100 000 to build a factory, which allows the country to create $100 000 dollars, then the moment you have borrowed it, you owe interest on it, and you may now end up having to pay back $180 000, but only $100 000 was created, where is the other $80 000 going to come from? It doesn't exit. What this does is lower the value of the money that has been created, leading to inflation and eventually economic collapse. So our economy, which is based on debt, is designed to collapse, from the beginning. Our planet generates its own energy, at its core, and there is almost unlimited power in the upper atmosphere, as a result. So why are we told that we need to burn fossil fuels and to have nuclear power stations?

Figure 18.4. A nuclear power station with steam pouring out of its cooling towers: Steam is necessary to produce rain, which would not otherwise occur, because of the pollution created by the burning of fossil fuels, as these produce CCNs (cloud condensation nuclei) are so small that the droplets of water, which form around them, are too small to ever fall as rain [1].

Why are we lied to in this manner? Why are we abused in this manner? Is it possible that we live in a world built on lies? Is there any historical record, which explains why this world is built on lies? Is there any Egyptian or Babylonian text that mentions this peculiarity about our world? No, no Egyptian or Babylonian, or Sumerian, or Chinese, text talks about the lies that seem to permeate life on this planet. There is only one that talks about this. This historical record is the Bible and this is what it states:

44 Ye are of your father the devil, and the lusts of your father ye will do. He was a murderer from the beginning, and abode not in the truth, because there is no truth in him. When he speaketh a lie, he speaketh of his own: for he is a liar and the father of it.

John 8:44

So, in the words of Jesus we have identified the Father of Lies, a spiritual being also called Satan, or Lucifer, and who in Genesis managed to get Adam and Eve to disobey and thus lose control of the planet. Thus, an alien being full of hatred for our species took over control of this planet. This is why we are lied to and abused at every turn. Our enemy rules this planet. He is not alone; there are many others under him who help him: the principalities and powers, the rulers of the darkness of this world also known as fallen angels or aliens. They too are liars; they have even lied to those human beings that serve them. They have told them that they will be able to leave the Solar System or that they can download their consciousness into a computer, and thus, be able to live forever. But the truth is that God has made sure that the Planet X System has enveloped the Solar System so that there is no way out (see Article 339: Planet X and the Interstellar Medium: can we leave the Solar System?) [6]. Evil has been contained inside the Solar System. And the only eternal life they will get unless they repent, is eternal life in the lake of fire. The AI technology or D wave computers are meant for the use of the demons, those beings who once had bodies, but were half human and half alien. These are also known as nephilim, and when they die they become demons. There are many alive, right now, masquerading as aliens, from other star systems. They are recognized by the fact that they cannot procreate normally, they have to be cloned because they are produced through genetic modification and just like genetically modified crops cannot procreate, through taking seed from a plant and growing it into another plant; these alien beings are the same way. All of these 'aliens' (nephilim) have the same nature as their fathers, they are liars.

Our enemy ruling this world is also a thief:

10 The thief does not come except to steal, and to kill, and to destroy. I have come that they may have life and that they may have it more abundantly.

John 10:10

Notice that that Jesus clearly places himself in opposition to this alien being. Jesus also faced him at the beginning of his ministry and after having a 40 day fast. This is one of the things that Lucifer tempted Jesus with:

8 Again, the devil took Him up on an exceedingly high mountain and showed Him all the kingdoms of the world and their glory. 9 And he said to Him, "All these things I will give You if You will fall down and worship me."

10 Then Jesus said to him, [b]"Away with you, Satan! For it is written, 'You shall worship the Lord your God, and Him only you shall serve.' "

Matt. 4:8 – 10

Notice that Lucifer said that he could give Jesus all the kingdoms of the world and Jesus did not refute it, so Lucifer was not lying about having control of the all the kingdoms of the world. He still does, and will, until Jesus returns to take over this world. This is our enemy. This is the one who tries to kill us and destroy us at every turn, whilst at the same time remaining hidden. Why does he do that? Because if we realized that he was real, we may just realize that God, the Creator, and His Son Jesus, are real as well.

One more question? Why did he allow us to see that the Sun was going dark during the SDO season when so many resources are used to hide what is going on with the Sun? Is it possible that he is compelled to show us the truth? Why would a liar show the truth? Well, he still lies, as he shows it and then declares that it is something else. In other words, he shows that the Sun is going dark but declares that it is only an eclipse. He then also makes sure that any astronomer or physicist who states what it really is that they are removed from their academic positions, or killed. Wouldn't it have been so much simpler and easier, for him, if the SDO eclipse season images were never allowed to be seen at all? Yes, it would, and the fact that he shows us the truth indicates that he is compelled to do so. Why? Is it possible that Lucifer is in competition with the Creator, and that he has to show the truth or otherwise he loses the challenge? And what would this challenge be about?

Is this why he wants to kill most of the world's population now, because there is a time limit to the challenge, and he has to get some things accomplished before the end, in order to try to win the challenge? Because if there was no time limit, then all he had to do was wait for the Planet X System to destroy the earth and all its inhabitants. But he needs to get human beings to worship him. That is what it is all about. That is what the challenge is. Who will we worship, Lucifer, or the true God, the Creator of the Universe? He is trying to give himself the advantage once again, He cannot force billions of people to worship him, but maybe he can manage to control a few 100 million.

So who are we? I think that Lucifer rebelled a long time ago and tried to invade Heaven, where God's throne is. Lucifer had one third of the angels under his control and so they followed him in the rebellion. The Bible says that his tail sweeps one third of the angels down to the Earth. We are those angels and God has given us a second chance to show that we are dedicated to Him, that we love Him and that we choose to worship Him. We get to choose here who we will worship. Lucifer though decided to give himself a great advantage, at the beginning of the challenge, and took over control of the planet and every human being is thus born as his slave. He is the prince of the power of the air, and so we hear his thoughts, which compel us to do what he wants. But he does not have complete control, and will not, unless we accept the mark of the beast, possibly an implanted chip, which will remove our power of free will.

It says in Romans 10:13: *For whosoever shall call upon the name of the Lord shall be saved*. If we call on the name of Jesus, or Yashua, if you prefer, as we recognize in our hearts that He is Lord, our Creator, and thus the one we belong to, we have just chosen Him.

So, choose wisely.

References:

[1]   Albers, C. (2018). Article 335: Biological and Ecological weapons in use against us.
[2]   http://archive.defense.gov/Transcripts/Transcript.aspx?TranscriptID=674
[3]   Albers, C. (2018). Article 244: The Planet X System: destroyer of Star Systems (in Book 7: Planet X The effects on the Earth and the Sun).
[4]   Albers, C. (2018). Article 210: Stellar Core gravity: tidal and G is not constant (in Book 6: Planet X Physicist Articles Part 1.
[5]   Albers, C. (2018). Article 288: Image of the center of the Milky Way Galaxy reveals how the universe works.
[6]   Albers, C. (2018). Article 324: Planet X causes the sun to be darkened.

Spectacular Observational Evidence from Planet X News

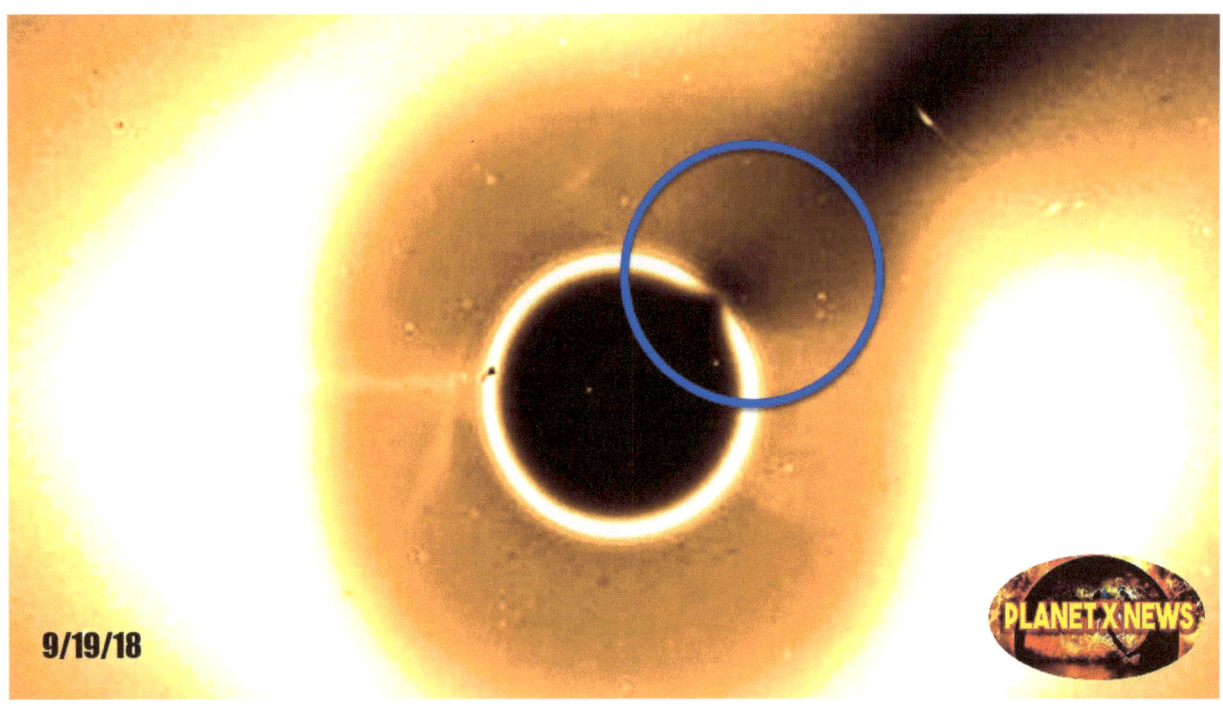

9/19/18

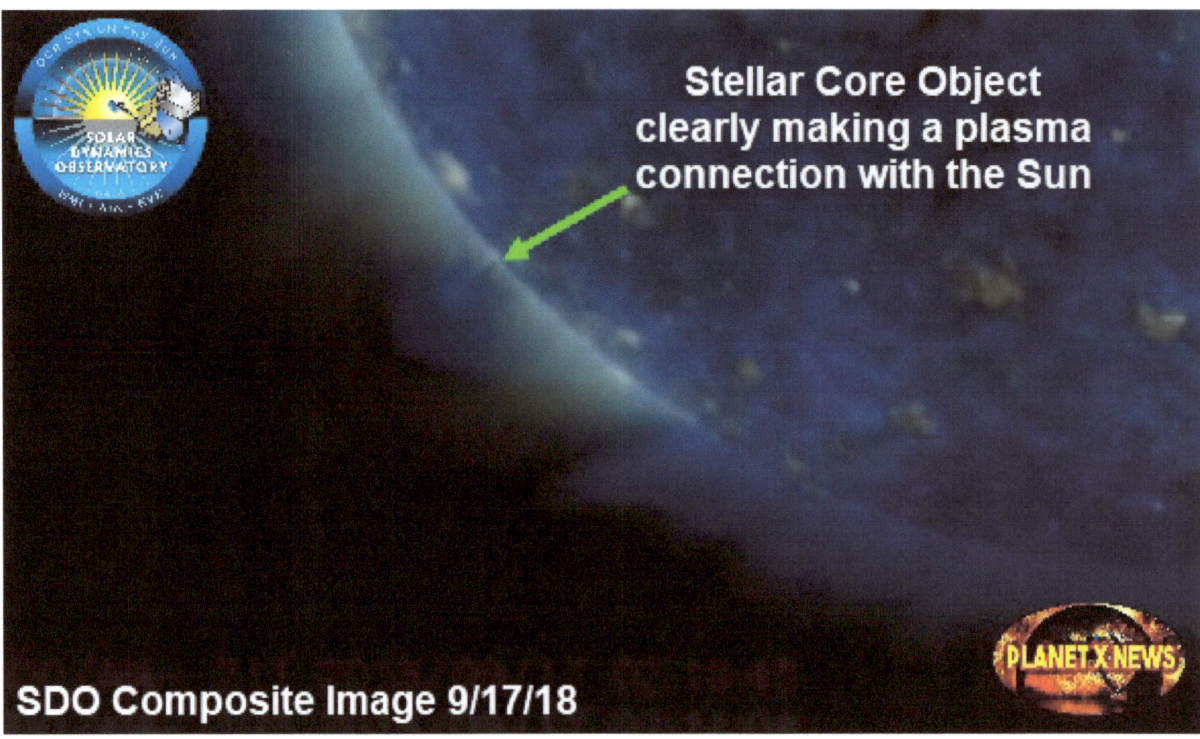

Stellar Core Object clearly making a plasma connection with the Sun

SDO Composite Image 9/17/18

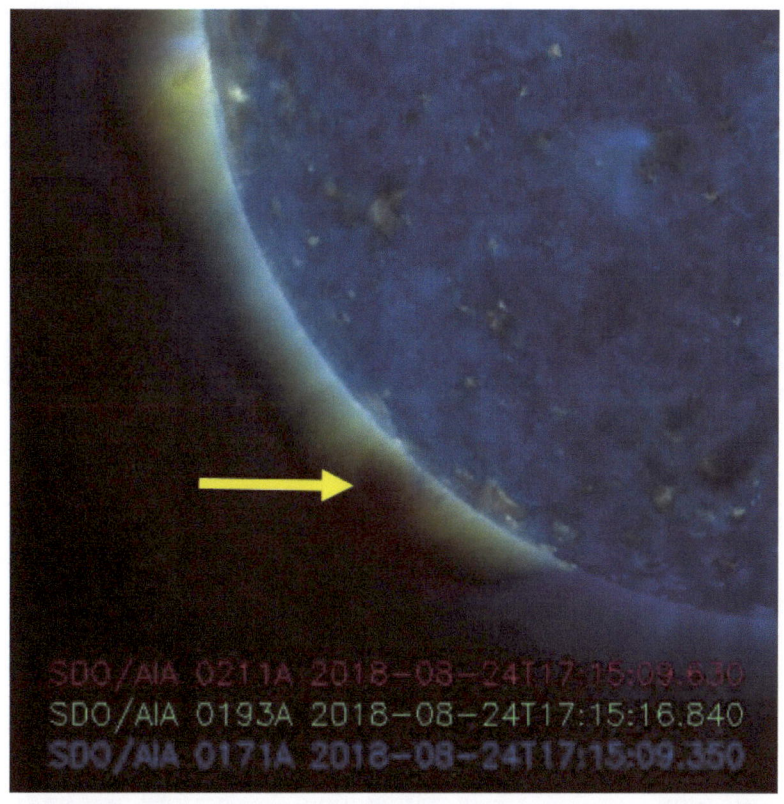
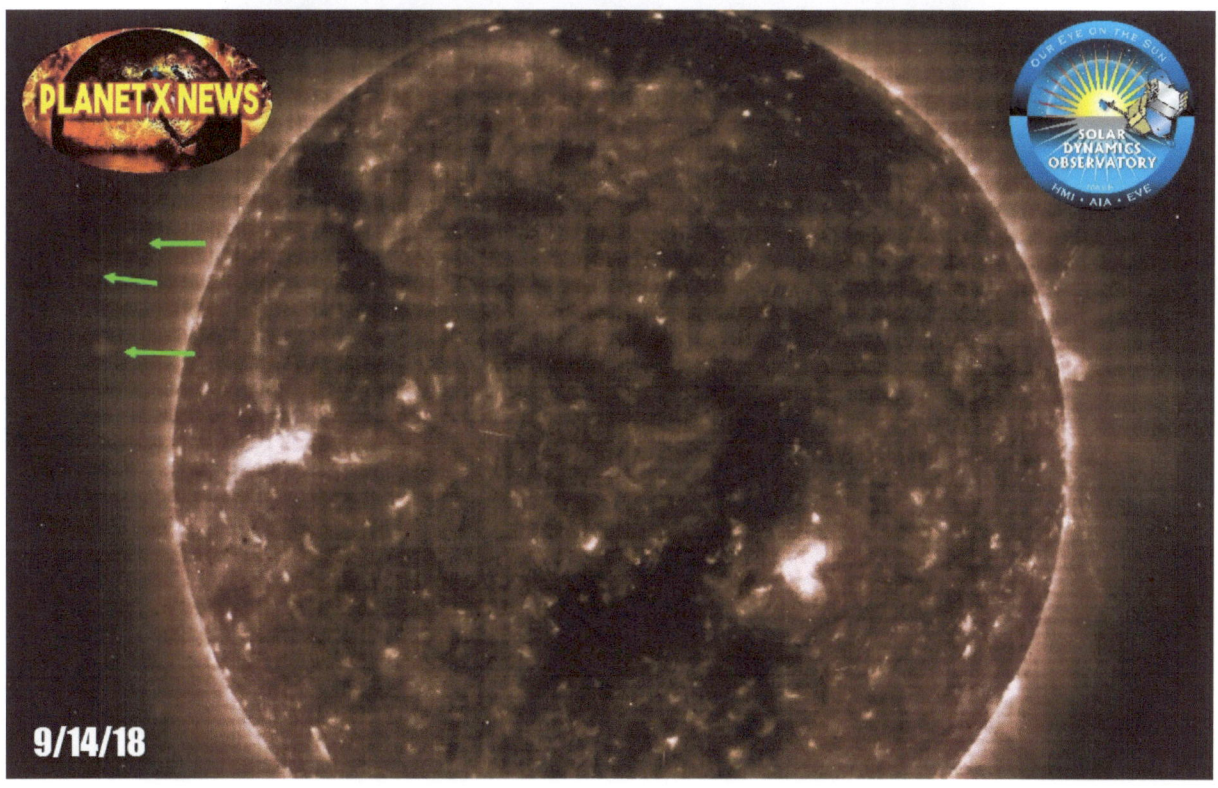

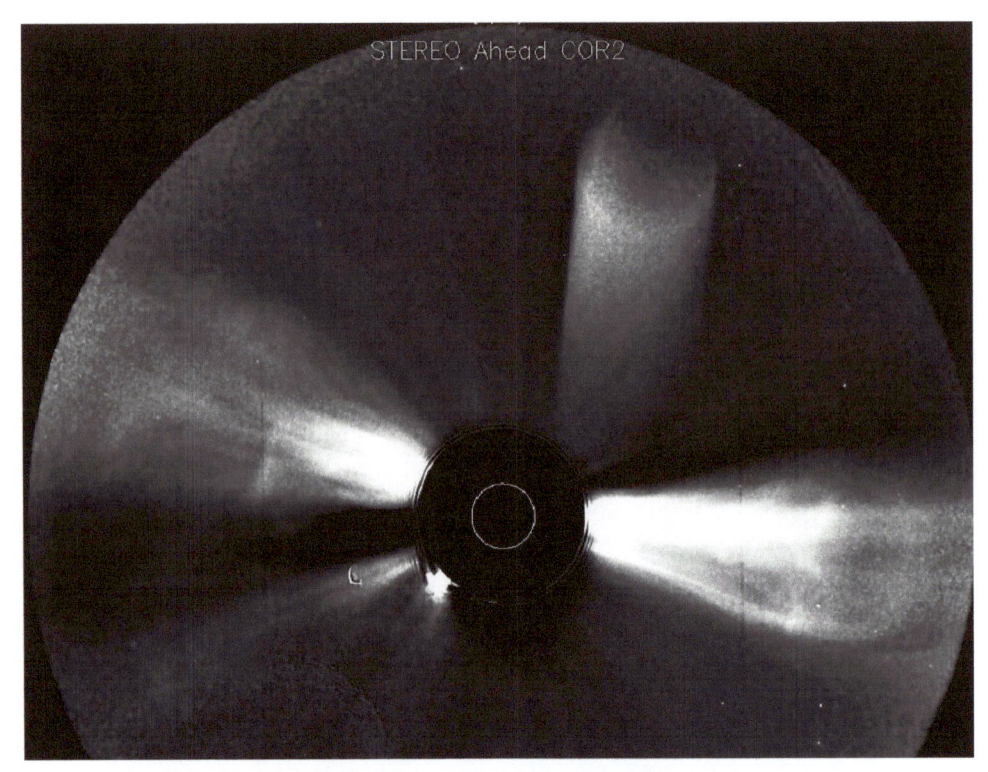

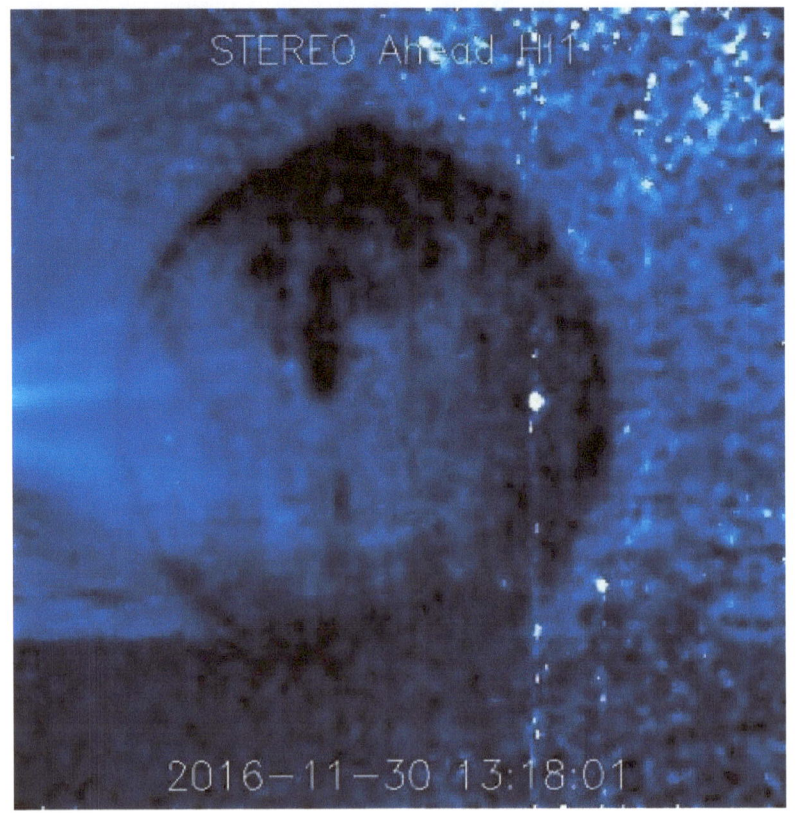

The End for Now!